A Beginner's Guide to Image Shape Feature Extraction Techniques

T0136408

Intelligent Signal Processing and Data Analysis

Series Editor

Nilanjan Dey, Department of Information Technology, Techno India College of Technology, Kolkata, India

Proposals for the series should be sent directly to one of the series editors above, or submitted to:

Chapman & Hall/CRC
Taylor and Francis Group
3 Park Square, Milton Park
Abingdon, OX14 4RN, UK

Bio-Inspired Algorithms in PID Controller Optimization
Jagatheesan Kaliannan, Anand Baskaran, Nilanjan Dey, and Amira S. Ashour

Digital Image Watermarking: Theoretical and Computational Advances
Surekha Borra, Rohit Thanki, and Nilanjan Dey

A Beginner's Guide to Image Preprocessing Techniques
Jyotismita Chaki and Nilanjan Dey

A Beginner's Guide to Image Shape Feature Extraction Techniques

Jyotismita Chaki
Vellore Institute of Technology

Nilanjan Dey
Techno India College of Technology

CRC Press
Taylor & Francis Group
Boca Raton London New York

CRC Press is an imprint of the
Taylor & Francis Group, an **Informa** business

MATLAB® is a trademark of The MathWorks, Inc. and is used with permission. The MathWorks does not warrant the accuracy of the text or exercises in this book. This book's use or discussion of MATLAB® software or related products does not constitute endorsement or sponsorship by The MathWorks of a particular pedagogical approach or particular use of the MATLAB® software.

CRC Press
Taylor & Francis Group
6000 Broken Sound Parkway NW, Suite 300
Boca Raton, FL 33487-2742

First issued in paperback 2021

© 2020 by Taylor & Francis Group, LLC
CRC Press is an imprint of Taylor & Francis Group, an Informa business

No claim to original U.S. Government works

Printed on acid-free paper

ISBN-13: 978-0-367-25439-1 (hbk)
ISBN-13: 978-1-03-209063-4 (pbk)

This book contains information obtained from authentic and highly regarded sources. Reasonable efforts have been made to publish reliable data and information, but the author and publisher cannot assume responsibility for the validity of all materials or the consequences of their use. The authors and publishers have attempted to trace the copyright holders of all material reproduced in this publication and apologize to copyright holders if permission to publish in this form has not been obtained. If any copyright material has not been acknowledged please write and let us know so we may rectify in any future reprint.

Except as permitted under U.S. Copyright Law, no part of this book may be reprinted, reproduced, transmitted, or utilized in any form by any electronic, mechanical, or other means, now known or hereafter invented, including photocopying, microfilming, and recording, or in any information storage or retrieval system, without written permission from the publishers.

For permission to photocopy or use material electronically from this work, please access www.copyright. com (http://www.copyright.com/) or contact the Copyright Clearance Center, Inc. (CCC), 222 Rosewood Drive, Danvers, MA 01923, 978-750-8400. CCC is a not-for-profit organization that provides licenses and registration for a variety of users. For organizations that have been granted a photocopy license by the CCC, a separate system of payment has been arranged.

Trademark Notice: Product or corporate names may be trademarks or registered trademarks, and are used only for identification and explanation without intent to infringe.

Visit the Taylor & Francis Web site at
http://www.taylorandfrancis.com

and the CRC Press Web site at
http://www.crcpress.com

Contents

Preface

Digital image processing is a widespread subject progressing continuously. Extracting image features has become a foremost player in many digital image processing applications. Shape feature detectors and descriptors have been explored and applied in several domains such as computer vision, pattern recognition, biometrics technology, image processing, medical image analysis, etc. Motivated by the requirement for an improved understanding of the shape feature detector fundamentals and application, this book presents different image shape extraction techniques which are essential for image retrieval. Shape feature extraction and depiction play a significant role in different types of applications such as: (1) Shape retrieval: searching for entire shapes in a huge database of shapes that are comparable to a query shape. Generally, all shapes in a specified distance from the query or the first limited shapes that have the minimum distance are calculated. (2) Shape recognition and classification: determining if a specified shape matches a model satisfactorily, or which of the database class is the most alike. (3) Shape alignment and registration: converting or interpreting one shape so that it best matches other shapes completely or in part. (4) Shape estimate and simplification: creating a shape of fewer elements (segments, points, etc.) that are still similar to the original. This book presents different techniques of one-dimensional, moment based, polygonal approximation based, spatial interrelation based, scale space based, transform shape domain-based shape feature extraction techniques in detail. The aim of this book is to not only present different perceptions of shape feature extraction techniques to undergraduate and postgraduate students but also to serve as a handbook for practicing engineers. Simulation is an important tool in any engineering field. In this book, the image shape extraction algorithms are simulated using MATLAB®. It has been the attempt of the authors to present detailed worked examples to demonstrate the various digital image shape feature extraction techniques.

The book is organized as follows:

Chapter 1 gives an overview of image shape feature. The importance of shape features used for image recognition are covered in this chapter. Different properties of shape features include translation, rotation and scale invariance, identifiability, noise resistance, affine invariance, occultation invariance, statistical independence and reliability are discussed in this chapter.

Chapter 2 deals with one-dimensional shape feature extraction techniques or shape signatures. The concept of complex coordinate is introduced in this chapter. This chapter also gives an overview of different one-dimensional shape feature extraction techniques such as centroid distance function, tangent angle, contour curvature, area function, triangle area representation and chord length distribution. The examples related to shape signatures are illustrated through MATLAB examples.

Chapter 3 is devoted to geometric shape feature extraction techniques. Different simple geometric shape feature extraction techniques such as center of gravity, axis of least inertia, digital bending energy, eccentricity, circularity ratio, elliptic variance, rectangularity, convexity, solidity, Euler number, profiles and hole area ratio are discussed step by step. Various eccentricity methods such as principal axis methods and minimum bounding rectangle methods are included in this chapter. The examples related to geometric shape feature extraction techniques are illustrated through MATLAB examples.

Chapter 4 discusses different polygonal approximation-based shape feature extraction techniques. Different polygonal approximation techniques such as merging, splitting, minimum perimeter polygon, dominant point detection, k-means method, genetic algorithm, ant colony optimization method and tabu search method are explained in this chapter. Various merging methods such as distance threshold method, tunneling method and polygon evolution are also included. The examples related to different polygonal approximation-based shape feature extraction techniques are illustrated through MATLAB examples.

The focus of Chapter 5 is on spatial interrelation-based shape feature extraction techniques. Different techniques such as adaptive grid resolution, bounding box, convex hull, chain code, smooth curve decomposition, ALI based representation, beam angle statistics, shape matrix, shape context, chord distribution and shock graph are explained in this chapter. Different methods to generate chain code such as basic, differential, resampling, vertex and chain code histograms are discussed in this chapter. Various techniques to create shape matrix like the square model and polar model are also explained step by step. The examples related to spatial interrelation-based shape feature extraction techniques are illustrated through MATLAB examples.

Chapter 6 provides an overview of moment-based shape feature extraction techniques. Different methods to extract moment shape feature including contour, geometric invariant, Zernike, radial Chebyshev, Legendre, homocentric polar-radius, orthogonal Fourier-Mellin and pseudo-Zernike are discussed in this chapter. The examples related to moment-based shape feature extraction techniques are illustrated through MATLAB examples.

Chapter 7 deals with scale space-based shape feature extraction techniques. Different methods including curvature scale space and intersection points map are discussed in this chapter. The examples related to scale space-based shape feature extraction techniques are illustrated through MATLAB examples.

The focus of Chapter 8 is on transform domain-based shape feature extraction techniques. Different techniques such as Fourier descriptors, wavelet transforms, angular radial transformation, shape signature harmonic embedding, $\Re$-transform and shapelet descriptors are discussed in this chapter. Types of Fourier descriptors including one-dimensional and region-based are also included in this chapter. The examples related to transform domain-based shape feature extraction techniques are illustrated through MATLAB examples.

Finally, Chapter 9 is devoted to various applications of shape features in pattern recognition in the areas of leaf recognition, fruit recognition, face recognition, hand gesture recognition, etc. The examples related to the application shape feature extraction techniques of color images are illustrated through MATLAB examples.

Dr. Jyotismita Chaki
Vellore Institute of Technology

Dr. Nilanjan Dey
Techno India College of Technology

MATLAB® is a registered trademark of The MathWorks, Inc. For product information, please contact:

The MathWorks, Inc.
3 Apple Hill Drive
Natick, MA 01760-2098 USA
Tel: 508 647 7000
Fax: 508-647-7001
E-mail: info@mathworks.com
Web: www.mathworks.com

Authors

Dr. Jyotismita Chaki is an assistant professor in the Department of Information Technology and Engineering in Vellore Institute of Technology, Vellore, India. She has done her PhD (Engg) in digital image processing from Jadavpur University, Kolkata, India. Her research interests include computer vision and image processing, pattern recognition, medical imaging, soft computing, data mining and machine learning. She has published one book and 22 international conferences and journal papers. She has also served as a Program Committee member of the *Second International Conference on Advanced Computing and Intelligent Engineering 2017 (ICACIE-2017)* and the *Fourth International Conference on Image Information Processing (ICIIP-2017)*.

Dr. Nilanjan Dey was born in Kolkata, India, in 1984. He received his BTech degree in Information Technology from West Bengal University of Technology in 2005, MTech in Information Technology in 2011 from the same university and PhD in digital image processing in 2015 from Jadavpur University, India. In 2011, he was appointed as an assistant professor in the Department of Information Technology at JIS College of Engineering, Kalyani, India, followed by Bengal College of Engineering College, Durgapur, India, in 2014. He is now employed as an assistant professor in Department of Information Technology, Techno India College of Technology, India. His research topics are signal processing, machine learning and information security. Dr. Dey is an associate editor of IEEE ACCESS and is currently the editor-in-chief of the *International Journal of Ambient Computing and Intelligence* and series editor of *Springer Tracts in Nature-Inspired Computing (STNIC)*.

1

Introduction to Shape Feature

1.1 Introduction

Visual information plays a significant role in our society and a progressively persistent role in our existence, and the need to preserve these sources is increasing. Images are utilized in several areas such as fashion, engineering design and architectural, advertising, entertainment, journalism, etc. Therefore, it delivers the needed prospect to utilize the richness of images [1]. The visual information will be impractical if it cannot be discovered. In the view of the practical and growing use of images, the capability to explore and to recover the images is a vital issue, necessitating image retrieval systems. Image visual features deliver an explanation of their content. Image retrieval (IR) occurred as a practical way to recover images and browse huge records of images. IR—the procedure of recovering images that is based on automatically extracted features—has been a subject of rigorous research in current years.

The input to a representative content based image retrieval (CBIR) and analysis system is a grey-scale image of a scene comprising the objects of interest. In order to understand the contents of a scene, it is essential to identify the objects positioned in the scene. The shape of the object is represented as a binary image which is the representative of the extent of the object. The shape can be assumed as a silhouette of the object (e.g., attained by revealing the object using a naturally distant light source). There are many imaging applications where image analysis can be minimized to the analysis of shapes, (e.g., machine parts, organs, characters, cells).

Shape analysis approaches analyze the objects in a scene. In this book, shape representation and explanation aspects of shape analysis are discussed. Shape representation approaches result in a non-numeric depiction of the original shape (e.g., a graph), so that the significant features of the shape are well-preserved. The word *significant* in the previous sentence typically has various meanings for various applications. The step following shape representation is shape description, which refers to the approaches that result in a numeric descriptor of the shape. A shape description technique produces a shape descriptor vector (also called a feature vector) from a specified shape [2]. The aim of description is to exclusively characterize the shape utilizing its shape descriptor vector. The input to shape analysis algorithms is shapes (i.e., binary images).

There are numerous techniques to obtain binary shape images from the grey-scale image (e.g., image segmentation). One of these techniques is connected component labeling. The neighborhood of a pixel is the group of pixels that touch it. Therefore, the neighborhood of a pixel can have a maximum of 8 pixels (images are always considered 2D). Figure 1.1 shows the neighbors (grey cells) of a pixel P.

There are different types of neighborhoods, described as follows.

1.1.1 4-Neighborhood

The 4-neighborhood contains only the pixels directly touching [3]. The pixel above, below, to the left, and right for the 4-neighborhood of a specific pixel as shown in Figure 1.2.

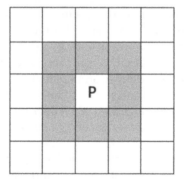

FIGURE 1.1
The grey pixels form the neighborhood of the pixel "P".

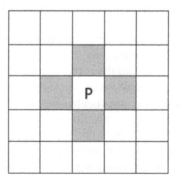

FIGURE 1.2
The grey pixels for the 4-neighbourhood of the pixel "P".

1.1.2 d-Neighborhood

This neighborhood is composed of those pixels that do not touch, or they touch the corners [4]. That is, the diagonal pixels as shown in Figure 1.3.

1.1.3 8-Neighborhood

This is the union of the 4-neighborhood and the d-neighborhood [5]. It is the maximum probable neighborhood that a pixel can have as shown in Figure 1.4.

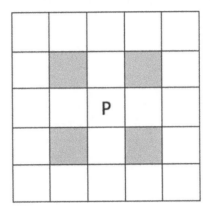

FIGURE 1.3
The diagonal neighborhood of the pixel "P" is shown in color.

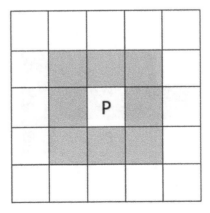

FIGURE 1.4
The 8-neighborhood.

1.1.4 Connectivity

Two pixels are supposed to be "connected" if they belong to the neighborhood of each other as shown in Figure 1.5.

The entire grey pixels are "connected" to "P" or they are 8-connected to "P". Therefore, only the dark grey ones are 4-connected to "P", and the light grey ones are d-connected to "P".

If there are several pixels, they are said to be connected if there is some "chain-of-connection" among any two pixels as shown in Figure 1.6.

Here, let's say the white pixels are considered the foreground set of pixels or the shape pixels. Then, pixels P_2 and P_3 are connected. There exists a chain of pixels which are connected to each other. However, pixels P_1 and P_2 are not connected. The black pixels (which are not in the foreground set of pixels) block the connectivity.

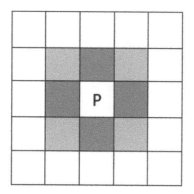

FIGURE 1.5
Connectivity of pixels.

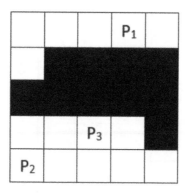

FIGURE 1.6
Multiple connectivity.

1.1.5 Connected Components

Taking the idea of connectivity, the idea of connected components is generated [6]. A graphic that best serves to clarify this idea is shown in Figure 1.7.

Figure 1.8 shows the connected component labeling example.

Each connected region has exactly one value (labeling).

Shapes can be represented in cartesian or polar coordinates. A cartesian coordinate (Figure 1.9a) is the simple x-y coordinate, and in polar coordinates (Figure 1.9b), shape elements are represented as pairs of angle θ and distance r.

FIGURE 1.7
Connected components.

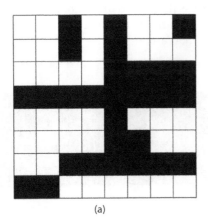

FIGURE 1.8
(a) Original image, (b) connected component labeling.

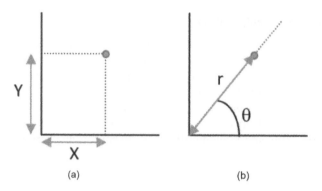

(a)   (b)

FIGURE 1.9
Coordinate system: (a) cartesian coordinate, and (b) polar coordinate.

1.2 Importance of Shape Features

Shape discrimination or matching refers to approaches for comparing shapes [7]. It is utilized in model-based object recognition where a group of known model objects is equated to an unidentified object spotted in the image. For this purpose, a shape description scheme is utilized to determine the shape descriptor vector for every model shape and unidentified shape in the scene. The unidentified shape is matched to one of the model shapes by equating the shape descriptor vectors utilizing a metric. In Figure 1.10b the grey-scale image of the original image is shown, in Figure 1.10c an overall shape of a leaf is presented, while boundary shape and interior details are illustrated in Figure 1.10d.

The basic idea by simple shape examples is illustrated in Figure 1.11. Shapes in Figure 1.11a–c are rotationally symmetric, and by this feature they can be differentiated from the shapes in Figure 1.11d and e.

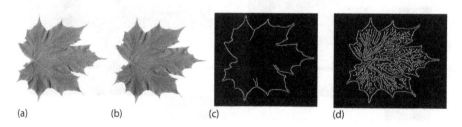

(a)   (b)   (c)   (d)

FIGURE 1.10
Shape representation. (a) Original image; (b) Grey-scale image; (c) Overall shape; (d) Boundary shape and interior details.

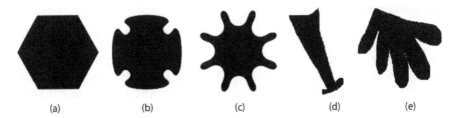

(a)   (b)   (c)   (d)   (e)

FIGURE 1.11
(a–e) Shape examples.

Assume a shape function S_1 is defined as a "symmetricity" measure. Additionally, since S_1 doesn't differentiate between the shapes in Figure 1.11a–c, another shape descriptor should be measured. Since the shape in Figure 1.11a is convex, and as the shape in Figure 1.11b is "more convex" than shape in Figure 1.11c, a "convexity" measure (i.e., another shape function) S_2 is considered. S_2 is probable to assign 1 to convex shapes and smaller values for "less convex" shapes (e.g., 0.92 to the shape in Figure 1.11b and 0.75 to the shape in Figure 1.11c). Such a demarcated function would differentiate among the shapes in Figure 1.11d and e, but it is not clear whether it is able to differentiate between shapes in Figure 1.11b and d. It is hard to judge which of them is "more convex". To overcome such a problem, another descriptor (e.g., shape "linearity") could be involved. A linearity measure S_3 should allocate a high value to the shape in Figure 1.11d (let say 0.9) and small, similar values for the rest of the shapes (e.g., all close to 0.1).

This is a fundamental process of using shape descriptors and equivalent measures to be able to differentiate between shapes/objects [8]. In most cases a single measure and a single descriptor are not sufficient, and sometimes they should be combined. For example, the shape in Figure 1.11e can be distinguished from the others as a shape with small S_1 and S_3 values (e.g., with low both symmetricity and linearity measures).

The extraction and depiction of shape features has a vital role in the following applications [9]:

- *Shape recovery*: finding the most similar shapes from a big database of comparable shapes to a query shape.
- *Classification and recognition of shape*: defining whether a specified shape satisfies a model or typical class that is the most alike.
- *Shape registration and alignment*: translating or converting one shape to match with another shape, in part or in whole.
- *Shape simplification and approximation*: creating a shape of smaller amounts of elements (triangles, points, segments, etc.) which is still equivalent to the original shape.

1.3 Properties of Efficient Shape Features

Shape descriptors can be used to efficiently discover similar shapes from the database even if they are affinely altered shapes like translated, rotated, flipped, scaled, etc [10]. The shape descriptor can also be used to efficiently search imperfect shapes, noise affected shapes, tolerated by human beings for the comparison and recovery of the shapes. This is identified as the robustness necessity. A shape descriptor should be able to accomplish image retrieval for maximum categories of shapes, not only for certain types of shapes. Hence, it should be application independent. One of the vital properties of the shape descriptor is low calculation complexity. By including fewer properties of the image in the calculation procedure, calculation can be lessened and lower computation complexity attained, and the shape descriptor becomes robust. Here low computation complexity means clarity and constancy.

Effective shape features must hold some important features as follows [11]:

- *Identifiability*: shapes that are alike by human perception must possess an identical feature which varies from others.

- *Rotation invariance*: the orientation of an object doesn't disturb its shape; consequently it would be probable that a shape descriptor should generate the identical measure for a shape S and for the similar shape oriented by θ degrees, $R(S, \theta)$.

- *Translation invariance*: the shape of an object is independent of the coordinate axes utilized; thus, it would be anticipated that a shape descriptor should generate a similar measure for the shape irrespective of its location in the coordinate plane.

- *Scale invariance*: since the shape of an object is independent of its depiction, the scale of an object should not affect the measure generated by a shape descriptor.

- *Affine invariance*: the affine change achieves a linear projection from a two-dimensional coordinate to other two-dimensional coordinates that conserves the parallelism and straightness of lines. Affine invariant features can be created utilizing arrangements of scales, orientation, translations, and shears.

- *Noise resistance*: features should be as reliable as probable against noise, i.e., they must be identical as noise resistance in a given range affecting the pattern.

- *Occultation invariance*: when certain shape portions are occulted by other shapes, the characteristics of the residual part must not alter equated to the original shape.

- *Statistically independent*: two characteristics of a shape must not be statistically dependent on each other. This characterizes the representation's compactness.
- *Reliable*: the extracted features must persist the similarity until one handles with the similar pattern.
- *Well-defined range*: having an awareness of the range of values formed by a shape descriptor can be significant when understanding the meaning of the values fashioned by the descriptor. Also, it may be beneficial to know the range fashioned by a descriptor in particular when designing an application.

Shape descriptor is generally some group of values that are formed to designate a specified feature of the shape. A descriptor attempts to measure a shape in such a manner that it is consistent with human perception. Good recognition rate of retrieval needs a shape feature that is able to efficiently find comparable shapes from a database [12]. The features are generally in the form of a vector. The shape feature should fulfill the subsequent needs:

- It should be complete enough to characterize the shape properly.
- It should be denoted and stored efficiently so that the descriptor vector size must not become very large.
- The calculation of distance among descriptors should be easy; otherwise a great deal of time is needed to implement it.

1.4 Types of Shape Features

Several shape explanation and depiction methods have been established for shape retrieval applications [13]. Based on whether the shape features are extracted from the contour only or are extracted from the whole shape region, the shape explanation and depiction methods are categorized into two categories:

- Contour-based approaches and
- Region-based approaches.

Every technique is further divided into two methods, structural approach and global approach. Structural methods and global methods are built on seeing if the shape is characterized by segments or as a whole.

1.4.1 Contour-Based Shape Representation and Description Techniques

By utilizing the contour-based methods, boundary or contour information will be extracted [14]. The contour-based shape representation method is further divided into global method and structural method. Global methods do not split the shape into subparts and the whole boundary information is utilized to originate the feature vector and for matching procedure, so this is also recognized as the continuous method. Structural methods divide the shape boundary information into segments or subparts called primitives, so this technique is also recognized as the discrete method. Usually the ultimate representation of the structural technique is a string or a graph (or tree), which will be utilized for matching for the image retrieval procedure.

1.4.1.1 Global Methods

From the shape contour information, a multi-dimensional numeric feature vector is produced that will be utilized for the matching procedure. Matching procedure is completed by manipulating the Euclidean distance or by point to point matching.

1.4.1.2 Structural Methods

In structural approaches of contour-based shape depiction and description method, the contour information is fragmented into segments, i.e., shapes are divided into boundary segments called primitives. The outcome is encoded into a string form like: $S = s_1, s_2, ..., s_n$; where, s_i may be an element of a shape feature which may contain a characteristic such as length, orientation, etc. The string can be directly utilized to signify the shape or can be utilized as an input to a recognition system for image retrieval procedure.

1.4.1.3 Limitations of the Structural Approach

Generation of primitives and features is the foremost limitation of the structural method. Since the number of primitives needed for every shape is not identified, there is no proper definition for an object or shape. The other limitation is the calculation of its complexity. This technique doesn't ensure for the best match. A variation of object contour causes alteration to the primitives, so it is more dependable than global approaches.

Contour-based approaches have an edge over region-based approaches in popularity because of the following reasons:

- Humans can effortlessly distinguish shapes by their contours.
- In several applications, the contour of a shape has importance, not its interior content.

These approaches also have some limitations:

- Contour-based shape descriptors are delicate to noise and alterations because of utilization of small parts of shapes.
- In several applications, contours are not obtainable.
- Some of the applications have more importance of their interior contents.

1.4.2 Region-Based Shape Representation and Description Techniques

Region-based approaches can overcome drawbacks of contour-based approaches [15]. They are more robust and can be useful in general applications. They can manage with shape defection. In the region-based method, all the pixels in the shape region, i.e., the entire region is considered for the shape representation and description. Comparable to the contour-based approaches, region-based approaches can also be divided into the global approaches and structural approaches, dependent on whether they divide the shapes into subparts or not.

Global approaches: Global approaches consider the entire shape region for shape representation and description.

Structural approaches: Region-based structural approaches divide the shape region into subparts that are utilized for shape representation and description. The region-based structural approaches have comparable problems to contour structural approaches.

The classification and some examples of shape descriptor is shown in Figure 1.12.

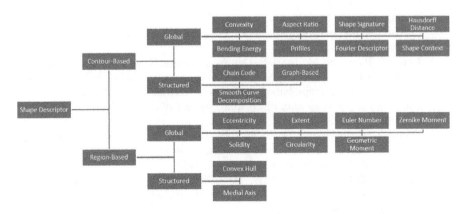

FIGURE 1.12
Classification and some examples of shape descriptor.

1.5 Summary

Defining the shape of an object can prove to be very difficult. Shape is generally characterized verbally or in figures. There is no commonly recognized procedure of shape description. Further, it is not identified what in the shape is significant. Current methods have both positive and negative qualities; computer graphics or mathematics utilize efficient shape representations that are impractical in shape recognition and vice versa. In spite of this, it is possible to discover features common to most shape description methods. Object description can be created on boundaries or on more complex knowledge of entire regions. Shape descriptors can be local as well as global. Global descriptors can only be utilized if whole object data are obtainable for analysis. Local descriptors designate local object properties utilizing partial information about the objects. Therefore, local descriptors can be utilized for description of occluded objects. Sensitivity to scale is even more serious if a shape description is derived, since shape may alter significantly with image resolution. Thus, shape should be considered in multiple resolutions which again can cause problems with matching corresponding shape representations from various resolutions. Furthermore, the conventional shape descriptions alter discontinuously.

References

1. He, L. F., Ren, X. W., Gao, Q. H., Zhao, X., Yao, B., & Chao, Y. Y. (2017, July). Fast basic shape feature computation. In *Computer Science and Artificial Intelligence-Proceedings of the International Conference on Computer Science and Artificial Intelligence (CSAI 2016)* (p. 18). World Scientific.
2. Chaki, J., Parekh, R., & Bhattacharya, S. (2015). Plant leaf recognition using texture and shape features with neural classifiers. *Pattern Recognition Letters, 58,* 61–68.
3. Araki, T., Ikeda, N., Molinari, F., Dey, N., Acharjee, S. M., Saba, L., Andrew, N., & Suri, J. S. (2014). Effect of geometric-based coronary calcium volume as a feature along with its shape-based attributes for cardiological risk prediction from low contrast intravascular ultrasound. *Journal of Medical Imaging and Health Informatics, 4*(2), 255–261.
4. Zheng, Y., & Doermann, D. (2006). Robust point matching for nonrigid shapes by preserving local neighborhood structures. *IEEE Transactions on Pattern Analysis and Machine Intelligence, 28*(4), 643–649.
5. Chaki, J. (2018). An efficient two-stage Palmprint recognition using Frangifilter and 2-component partition method. In *2018 Fifth International Conference on Emerging Applications of Information Technology (EAIT)* (pp. 1–5). IEEE.

6. Maji, P., Chatterjee, S., Chakraborty, S., Kausar, N., Samanta, S., & Dey, N. (2015). Effect of Euler number as a feature in gender recognition system from offline handwritten signature using neural networks. In *2015 2nd International Conference on Computing for Sustainable Global Development* (*INDIACom*) (pp. 1869–1873). IEEE.

7. Chaki, J., Parekh, R., & Bhattacharya, S. (2015). Recognition of whole and deformed plant leaves using statistical shape features and neuro-fuzzy classifier. In *2015 IEEE 2nd International Conference on Recent Trends in Information Systems* (*ReTIS*) (pp. 189–194). IEEE.

8. Tharwat, A., Gaber, T., Awad, Y. M., Dey, N., & Hassanien, A. E. (2016). Plants identification using feature fusion technique and bagging classifier. In *The 1st International Conference on Advanced Intelligent System and Informatics* (*AISI2015*), *November 28–30, 2015, Beni Suef, Egypt* (pp. 461–471). Springer, Cham, Switzerland.

9. Chaki, J., Parekh, R., & Bhattacharya, S. (2016). Plant leaf recognition using ridge filter and curvelet transform with neuro-fuzzy classifier. In *Proceedings of 3rd International Conference on Advanced Computing, Networking and Informatics* (pp. 37–44). Springer, New Delhi, India.

10. Virmani, J., Dey, N., & Kumar, V. (2016). PCA-PNN and PCA-SVM based CAD systems for breast density classification. In *Applications of Intelligent Optimization in Biology and Medicine* (pp. 159–180). Springer, Cham, Switzerland.

11. Chaki, J., Parekh, R., & Bhattacharya, S. (2016). Plant leaf recognition using a layered approach. In *2016 International Conference on Microelectronics, Computing and Communications* (*MicroCom*) (pp. 1–6). IEEE.

12. Tian, Z., Dey, N., Ashour, A. S., McCauley, P., & Shi, F. (2017). Morphological segmenting and neighborhood pixel-based locality preserving projection on brain fMRI dataset for semantic feature extraction: An affective computing study. *Neural Computing and Applications*, 1–16.

13. Chaki, J., Parekh, R., & Bhattacharya, S. (2018). Plant leaf classification using multiple descriptors: A hierarchical approach. *Journal of King Saud University-Computer and Information Sciences*.

14. AlShahrani, A. M., Al-Abadi, M. A., Al-Malki, A. S., Ashour, A. S., & Dey, N. (2018). Automated system for crops recognition and classification. In *Computer Vision: Concepts, Methodologies, Tools, and Applications* (pp. 1208–1223). IGI Global, Hershey, PA.

15. Chaki, J., & Parekh, R. (2012). Designing an automated system for plant leaf recognition. *International Journal of Advances in Engineering & Technology*, 2(1), 149.

2

One-Dimensional Function Shape Features

The one-dimensional shape feature function that is obtained from edge coordinates of the shape is also frequently called the shape signature. The shape signature generally holds the perspective shape property of the object [1]. Shape signature can define the entire shape; it is also frequently utilized as a pre-processing to other feature extraction procedures. In this chapter, the shape signatures or one-dimensional shape feature functions are presented.

2.1 Complex Coordinate (ComC)

A complex coordinate function is basically the complex number obtained from the contour point coordinates $B_n(x(n), y(n))$, $n \in [1, L]$ of an object and can be represented by equation (2.1) [2].

$$CC(n) = x(n) + iy(n) \qquad (2.1)$$

There are some advantages and limitations of ComC that are presented here.

Advantages: The advantage of using the complex coordinate function is it involves no extra computation in deriving shape signature and is also invariant to translation.

Limitations: The method is not invariant to translation. To make the shape representation of equation (2.1) invariant to translation, the equation is represented as shown in equation (2.2).

$$CC(n) = \left[x(n) - x_0\right] + i\left[y(n) - y_0\right] \qquad (2.2)$$

where $i = \sqrt{-1}$ and (x_0, y_0) is the centroid of the shape and can be presented by equation (2.3).

$$x_0 = \frac{1}{L}\sum_{n=0}^{L-1} x(n); y_0 = \frac{1}{L}\sum_{n=0}^{L-1} y(n); \qquad (2.3)$$

where L is the number of overall contour points.

2.2 Centroid Distance Function (CDF)

CDF is articulated by the distance of the contour points from the centroid (x_0, y_0) of a shape and is represented by equation (2.4) [3].

$$r(n) = \sqrt{\left(x(n) - x_0\right)^2 + \left(y(n) - y_0\right)^2} \tag{2.4}$$

The centroid is located at the position (x_0, y_0) which are, respectively, the average of the x and y coordinates for all contour points. The boundary of a shape consists of a series of contour or boundary points. A radius is a straight line joining the centroid to a boundary point. In the centroid distance model, lengths of a shape's radii from its centroid at regular intervals are captured as the shape's descriptor using the Euclidean distance [4]. More formally, let θ be the regular interval (measured in degrees) between radii (Figure 2.1). Then, the number of intervals is given by $k = 360/\theta$. All radii lengths are normalized by dividing with the longest radius length from the set of radii lengths extracted. Furthermore, without loss of generality, suppose that the intervals are taken clockwise starting from the x-axis direction. Then, the shape descriptor can be represented as a vector as shown in equation (2.5).

$$S = \left\{r_0, r_\theta, r_{2\theta}, ..., r_{(k-1)\theta}\right\} \tag{2.5}$$

Here $r_{i\theta}$, $0 \leq i \leq (k-1)$ is the $(i+1)$-th radius from the centroid to the boundary of the shape. With sufficient number of radii, dissimilar shapes can be differentiated from each other [5].

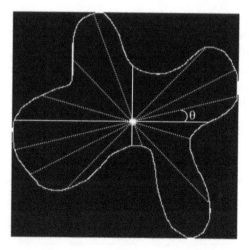

FIGURE 2.1
Centroid distance approach.

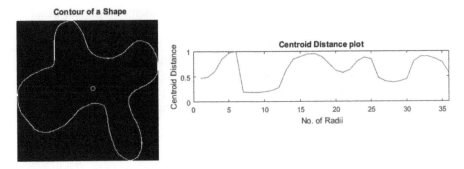

FIGURE 2.2
Considering $\theta = 10$ centroid distance plot of 36 radii.

Figure 2.2 shows the centroid distance plot of a shape boundary.
There are some advantages and limitations of this method that are presented here.

Advantages: Because of the deduction of centroid that designates the position of the shape from edge coordinates, this method is invariant to translation.

Limitations: If there is more than one boundary point at same interval, this method fails to represent the shape properly.

2.3 Tangent Angle (TA)

TA at a point $B_n(x(n), y(n))$ is demarcated by a tangential direction (θ) of a boundary as represented in equation (2.6) [6].

$$\theta(n) = \theta_n = \arctan\frac{y(n) - y(n - P)}{x(n) - x(n - P)} \qquad (2.6)$$

As each boundary is a digital curve, P is a small window to compute $\theta(n)$ more correctly.
There are some advantages and limitations of this method that are presented here.

Advantages: Tangent angle function is capable of reconstructing the original data.

Limitations: This method has two issues [7]. One is sensitivity to noise. To reduce the noise effect, a low-pass filter with suitable bandwidth is used to filter the contour prior to using this method. The other limitation is discontinuity because of the fact that this method undertakes values generally in the interval of $[-\pi, \pi]$ or $[0, 2\pi]$. Thus, in general θ_n comprises discontinuities of size 2π. To get rid from the discontinuity issue, with a random initial point, the cumulative angular function φ_n is demarcated as the angle variances among the tangent at any curve point P_n and the tangent at the initial point P_0 as shown in equation (2.7).

$$\varphi_n = \left[\theta_n - \theta_0\right] \tag{2.7}$$

Considering the fact that in human perception a circle is shapeless, let $t = 2\pi n/L$, then $\varphi_n = \varphi_{tL/2\pi}$. A periodic function is called the cumulative angular deviant function $\psi(t)$ and is demarcated as shown in equation (2.8).

$$\psi_t = \varphi_{(tL/2\pi)-t}; t \in \left[0, 2\pi\right] \tag{2.8}$$

L is the number of entire points in the contour.

2.4 Contour Curvature (CC)

CC is a very significant contour characteristic to measure the likeness among shapes [8]. It also has salient perceptual features and is confirmed to be very beneficial for recognition of the shape. With the aim of using $C(n)$ for shape depiction, the function of curvature $C(n)$ is represented by equation (2.9).

$$C(n) = \frac{\dot{x}(n)\ddot{y}(n) - \dot{y}(n)\ddot{x}(n)}{\left(\dot{x}(n)^2 + \dot{y}(n)^2\right)^{3/2}} \tag{2.9}$$

where $\dot{x}, \ddot{x}$ are the single and double derivative (gradient) in x-direction and $\dot{y}, \ddot{y}$ are the single and double derivative (gradient) in y-direction.

Thus, it is possible to calculate the planar curve curvature from its parametric depiction. If n is represented by the standardized arc-length parameter s, then equation (2.9) can be represented as shown in equation (2.10).

$$C(s) = \dot{x}(s)\ddot{y}(s) - \dot{y}(s)\ddot{x}(s) \tag{2.10}$$

As specified in equation (2.9), CC is calculated only from parametric derivatives, and, thus, it is invariant to translations and orientations [9]. CC is proportional to the scale in inverse. A probable technique to attain scale invariance is to standardize CC by the average absolute curvature, as represented in equation (2.11).

$$C'(s) = \frac{C(s)}{\frac{1}{L}\sum_{s=1}^{L}|C(s)|} \tag{2.11}$$

where L is the number of points residing on the standardized contour.

When the curve size is a vital characteristic for discrimination, CC should be utilized without the standardization; for the determination of shape analysis which is invariant to scale, the standardization should be achieved.

Initiating from a random point and succeeding the contour in clockwise direction, the curvature at every interpolated point is computed utilizing equation (2.10). Concave and convex vertices will infer positive and negative values, correspondingly (the opposite is substantiated for counter clockwise sense) [10].

Figure 2.3 is an example of CC. Clearly, a CC descriptor can discriminate various shapes.

There are some advantages and limitations of this method that are presented here.

Advantages: The main advantage of this method is that it is steady with respect to scale, noise, and change in orientation. It is an information-preserving feature and possesses an intuitively pleasing correspondence to the perceptive property of simplicity.

Limitations: Minor change in shape can affect the contour curvature largely.

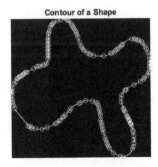
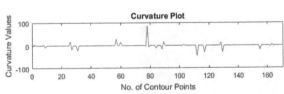

FIGURE 2.3
Curvature plot of a shape (177 contour points to generate the curvature are indicated with small circles).

2.5 Area Function (AF)

When the contour points are altered along the shape edge, the area of the triangle modeled by two consecutive contour points and the centroid also alter [11]. This produces an area function that can be considered a shape depiction. Figure 2.4 demonstrates this. Let A_n be the area among the consecutive edge points P_n, P_{n+1} and centroid C.

Figure 2.5 shows the area function plot of a shape boundary.

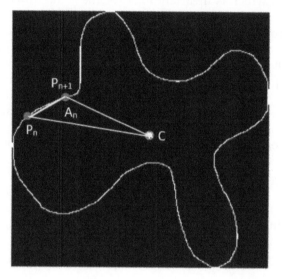

FIGURE 2.4
Area function approach.

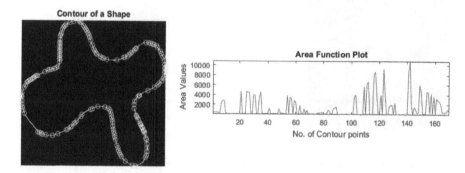

FIGURE 2.5
Area plot of a shape (170 contour points to generate the plot are indicated with small circles).

There are some advantages and limitations of AF that are presented here.

Advantages: This method is linear under affine change. This method is rotation invariant up to parametrization with dense enough boundary sampling and is also translation invariant.

Limitations: The linearity under affine transform only works for a shape sampled at its identical vertices.

2.6 Triangle Area Representation (TAR)

TAR [12] signature is calculated from the area of the triangles created by the contour points on the shape. The curvature of the boundary point (x_n, y_n) is computed utilizing TAR as follows.

For every three successive points $B_{n - Ps} (x_{n - Ps}, y_{n - Ps})$, $B_n(x_n, y_n)$, and $B_{n + Ps} (x_{n + Ps}, y_{n + Ps})$, where $n \in [1, L]$ and $P_s \in [1, L/2 - 1]$, L indicates even contour points. The signed area of the triangle created by these points is specified by equation (2.12).

$$TAR(n, P_s) = \frac{1}{2} \begin{vmatrix} x_{n-P_s} & y_{n-P_s} & 1 \\ x_n & y_n & 1 \\ x_{n+P_s} & y_{n+P_s} & 1 \end{vmatrix} \qquad (2.12)$$

When the boundary is passed through in anti-clockwise direction, zero, negative, and positive values of TAR represent straight-line, concave, and convex points, respectively [13].

Figure 2.6 establishes the entire TAR signature for the star shape.

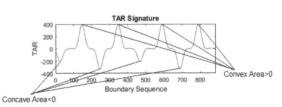

FIGURE 2.6
TAR signature of a shape.

By growing the triangle side lengths, i.e., taking further points, equation (2.12) will characterize longer discrepancies along the edge. The TARs with various triangle sides can be denoted as various scale space functions. The entire TARs, $P_s \in [1, L/2 - 1]$, comprise a multi-scale space TAR.

There are some advantages and limitations of this method that are presented here.

Advantages: TAR is affine-invariant, vigorous to noise, and robust to rigid transformation.

Limitations: TAR has a high computational cost as the most contour points are used [14]. Furthermore, TAR has two main limitations: (1) The area is not revealing about the type of the triangle (equilateral, isosceles, etc.) considered, which may be vital for a local description of the boundary. (2) The area is not sufficiently precise to characterize the shape of a triangle.

2.7 Chord Length Function (CLF)

CLF is obtained from shape contour without utilizing any reference point [15]. For every contour point C, its CLF is the minimum distance among C and the other contour point C' so that line CC' is orthogonal to the tangent vector at C (Figure 2.7).

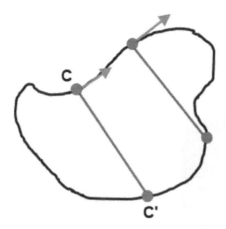

FIGURE 2.7
Chord length function approach.

There are some advantages and limitations of this method that are presented here.

Advantages: This method is not variant to translation and it controls the biased reference point (the fact that the centroid is frequently biased by contour defections or noise) issues.

Limitations: The chord length function is very delicate to noise, and in the signature of even smoothed shape boundaries can cause an extreme burst.

2.8 Summary

A shape signature describes a shape with a 1D function from the shape edge. To attain the translation invariance, it is generally demarcated by relative values. Normalization is essential to achieve invariance in scale. To recompense for rotation variations, shift matching is desirable to search the most similar among two shapes. TA, CC, and TAR have the invariance property in terms of occultation. Furthermore, computing shape signature from a shape is very easy. Shape signatures are delicate to noise, and small variations in the edge can result in huge errors while shape matching. Hence, some additional processing is required to raise its strength and decrease the similarity checking load. For instance, a shape signature can be refined by quantizing the rotationally invariant signature histogram.

References

1. Zhang, D., & Lu, G. (2004). Review of shape representation and description techniques. *Pattern Recognition, 37*(1), 1–19.
2. Zhao, X., Cheng, X., Zhou, J., Xu, Z., Dey, N., Ashour, A. S., & Satapathy, S. C. (2018). Advanced topological map matching algorithm based on D–S theory. *Arabian Journal for Science and Engineering, 43*(8), 3863–3874.
3. Chaki, J., & Parekh, R. (2011). Plant leaf recognition using shape based features and neural network classifiers. *International Journal of Advanced Computer Science and Applications, 2*(10), 41–47.
4. Dey, N., Roy, A. B., Pal, M., & Das, A. (2012). FCM based blood vessel segmentation method for retinal images. *arXiv preprint arXiv:1209.1181*.
5. Chaki, J., & Parekh, R. (2012). Designing an automated system for plant leaf recognition. *International Journal of Advances in Engineering & Technology, 2*(1), 149.

6. Li, Z., Shi, K., Dey, N., Ashour, A. S., Wang, D., Balas, V. E., McCauley, P., & Shi, F. (2017). Rule-based back propagation neural networks for various precision rough set presented KANSEI knowledge prediction: A case study on shoe product form features extraction. *Neural Computing and Applications, 28*(3), 613–630.

7. Boscaini, D., Masci, J., Rodolà, E., & Bronstein, M. (2016). Learning shape correspondence with anisotropic convolutional neural networks. In *Advances in Neural Information Processing Systems* (pp. 3189–3197).

8. Kumar, R., Talukdar, F. A., Dey, N., Ashour, A. S., Santhi, V., Balas, V. E., & Shi, F. (2017). Histogram thresholding in image segmentation: A joint level set method and lattice boltzmann method based approach. In *Information Technology and Intelligent Transportation Systems* (pp. 529–539). Springer, Cham, Switzerland.

9. Chaki, J., & Dey, N. (2018). *A Beginner's Guide to Image Preprocessing Techniques.* CRC Press, Boca Raton, FL.

10. Yang, J., Wang, H., Yuan, J., Li, Y., & Liu, J. (2016). Invariant multi-scale descriptor for shape representation, matching and retrieval. *Computer Vision and Image Understanding, 145*, 43–58.

11. Dey, N., Das, P., Roy, A. B., Das, A., & Chaudhuri, S. S. (2012, October). DWT-DCT-SVD based intravascular ultrasound video watermarking. In *2012 World Congress on Information and Communication Technologies (WICT 2012)* (pp. 224–229). IEEE.

12. Zhang, D., & Lu, G. (2002, January). A comparative study of Fourier descriptors for shape representation and retrieval. In *Proceedings of the 5th Asian Conference on Computer Vision* (p. 35). Springer.

13. Kamal, M. S., Nimmy, S. F., Hossain, M. I., Dey, N., Ashour, A. S., & Santhi, V. (2016, March). ExSep: An exon separation process using neural skyline filter. In *2016 International Conference on Electrical, Electronics, and Optimization Techniques (ICEEOT)* (pp. 48–53). IEEE.

14. Wang, D., Li, Z., Dey, N., Ashour, A. S., Moraru, L., Biswas, A., & Shi, F. (2019). Optical pressure sensors based plantar image segmenting using an improved fully convolutional network. *Optik, 179*, 99–114.

15. Kulfan, B. M. (2008). Universal parametric geometry representation method. *Journal of Aircraft, 45*(1), 142–158.

3

Geometric Shape Features

Shape-based image recovery consists of determining likeness among shapes characterized by their features. Some simple and basic geometrical characteristics can be utilized to designate shapes [1]. It distinguishes the shape of an image by utilizing geometrical features and then chooses the cluster of images that match that shape from a huge database.

3.1 Center of Gravity (CoG)

CoG is also known as centroid [2]. Its location should be static relative to the shape. If a shape is signified by its region function, its centroid (R_x, R_y) is represented by equation (3.1).

$$R_x = \frac{1}{M} \sum_{i=1}^{M} x_i$$
$$R_y = \frac{1}{M} \sum_{i=1}^{M} y_i \tag{3.1}$$

where M is the number of pixels or points in the shape.

Figure 3.1 illustrates the centroid of a binary region which is demarcated by a circle.

If a shape is denoted by its edge, the location of its centroid (C_x, C_y) is represented by equation (3.2).

$$C_x = \frac{1}{6A} \sum_{i=0}^{M-1} (x_i + x_{i+1})(x_i y_{i+1} - x_{i+1} y_i)$$
$$C_y = \frac{1}{6A} \sum_{i=0}^{M-1} (y_i + y_{i+1})(x_i y_{i+1} - x_{i+1} y_i) \tag{3.2}$$

where A is the contour's area and can be represented by equation (3.3).

$$A = \frac{1}{2} \left| \sum_{i=0}^{M-1} (x_i y_{i+1} - x_{i+1} y_i) \right| \tag{3.3}$$

FIGURE 3.1
Centroid of a binary region.

FIGURE 3.2
Centroid of a binary contour.

Figure 3.2 illustrates the centroid of a binary contour which is demarcated by a circle.

There are some advantages and limitations of this method that are presented here.

Advantages: The shape centroid location is static with various contour points distribution. The centroid position is static regardless of the spreading of points.

Limitations: This method usually can only discriminate shapes with large variations. This method is not suitable for standalone shape descriptors.

3.2 Axis of Minimum Inertia (AMI)

AMI is exclusive to the shape [3]. It assists as an exclusive reference line to reserve the shape rotation. AMI of a shape is demarcated by the smallest line calculated by the integral of the square of the distances to points on the contour of the shape. Since AMI goes through the contour centroid, to discover the AMI, transfer the shape and permit the shape centroid be the Cartesian coordinate system's origin. Let the parametric equation of AMI be denoted as $x \sin\alpha - y \cos\alpha = 0$. The angle of the slope is α.

Let φ be the angle amid the AMI and the x-axis. The inertia is represented by equation (3.4).

$$I = \frac{1}{2}(p+r) - \frac{1}{2}(p-r)\cos(2\varphi) - \frac{1}{2}q\sin(2\varphi)$$

$$\text{where}: p = \sum_{i=0}^{M-1} x_i^2; q = 2\sum_{i=0}^{M-1} x_i y_i; r = \sum_{i=0}^{M-1} y_i^2$$

(3.4)

Thus,

$$\frac{dI}{d\varphi} = (p-r)\sin(2\varphi) - q\cos(2\varphi)$$

$$\frac{d^2I}{d\varphi^2} = 2(p-r)\cos(2\varphi) + 2q\sin(2\varphi)$$

Let $dI/d\varphi = 0$, and φ can be represented by equation (3.5).

$$\varphi = \frac{1}{2}\arctan\left(\frac{q}{p-r}\right), -\frac{\pi}{2} < \varphi < \frac{\pi}{2}$$

(3.5)

α is assessed by using equation (3.6).

$$\alpha = \begin{cases} \varphi + \dfrac{\pi}{2} & if : \dfrac{d^2I}{d\varphi^2} < 0 \\ \varphi & \text{otherwise} \end{cases}$$

(3.6)

There are some advantages and limitations of this method that are presented here.

Advantages: This method is unique for every shape. This representation is invariant to orientation, scaling, flipping (reflection), and translation.

Limitations: Choosing the proper number of features for a particular shape is difficult. The coordinate of the object needs to be transferred to a Cartesian coordinate system. This method is not suitable for standalone shape descriptors.

3.3 Average Bending Energy (ABE)

ABE [4] is illustrated by equation (3.7).

$$\text{ABE} = \frac{1}{M} \sum_{t=0}^{M-1} K(t)^2 \tag{3.7}$$

where curvature function is $K(t)$, t is the arc length parameter, and M is the number of contour points. The circle has the least average bending energy.

Figure 3.3 shows the average bending energy of some shapes.

There are some advantages and limitations of this method that are presented here.

Advantages: This representation is rotationally invariant.

Limitations: This method usually can only discriminate shapes with large variations. This method is not suitable for standalone shape descriptors.

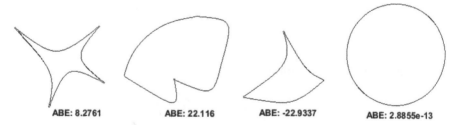

ABE: 8.2761   ABE: 22.116   ABE: -22.9337   ABE: 2.8855e-13

FIGURE 3.3
Average bending energy of some shapes.

3.4 Eccentricity

Eccentricity is the calculation of aspect ratio. It is basically the major axis length to the minor axis length ratio, thus, though the image is rotated, scaled, or translated, eccentricity remains identical [5]. Eccentricity returns a scalar that denotes the eccentricity of the ellipse that has the same second-moments as the region. The eccentricity is the ratio of the distance among the foci of the ellipse and its major axis length. The value is between 0 and 1. (0 and 1 are degenerate cases. An ellipse whose eccentricity is 0 is actually a circle, while an ellipse whose eccentricity is 1 is a line segment.) It can be computed by minimum bounding rectangle or principal axes technique.

3.4.1 Principal Axes Method

Principal axes of a certain shape can be exclusively demarcated as two-line segments that cross perpendicularly in the shape's centroid and denote the zero cross-correlation directions [6]. By this technique, a border is perceived as a statistical distribution occurrence. The contour covariance matrix C is represented by equation (3.8).

$$C = \frac{1}{M} \sum_{i=0}^{M-1} \begin{pmatrix} x_i - R_x \\ y_i - R_y \end{pmatrix} \begin{pmatrix} x_i - R_x \\ y_i - R_y \end{pmatrix}^T = \begin{pmatrix} c_{xx} & c_{xy} \\ c_{yx} & c_{yy} \end{pmatrix}$$

where:

$$c_{xx} = \frac{1}{M} \sum_{i=0}^{M-1} (x_i - R_x)^2$$

$$c_{xy} = \frac{1}{M} \sum_{i=0}^{M-1} (x_i - R_x)(y_i - R_y)$$

$$c_{yx} = \frac{1}{M} \sum_{i=0}^{M-1} (y_i - R_y)(x_i - R_x)$$

$$c_{yy} = \frac{1}{M} \sum_{i=0}^{M-1} (y_i - R_y)^2$$

(3.8)

Here (R_x, R_y) is the shape centroid and $c_{xy} = c_{yx}$.

The two principal axes lengths are identical with the eigenvalues λ_1 and λ_2 of the contour covariance matrix C, correspondingly, which can be represented by equation (3.9).

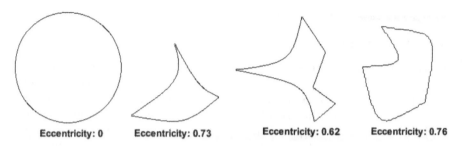

Eccentricity: 0   **Eccentricity: 0.73**   **Eccentricity: 0.62**   **Eccentricity: 0.76**

FIGURE 3.4
Eccentricity of some shapes.

$$\lambda_1 = \frac{1}{2}\left[c_{xx} + c_{yy} + \sqrt{\left(c_{xx} + c_{yy}\right)^2 - 4\left(c_{xx}c_{yy} - c_{xy}^2\right)} \right]$$
$$\lambda_2 = \frac{1}{2}\left[c_{xx} + c_{yy} - \sqrt{\left(c_{xx} + c_{yy}\right)^2 - 4\left(c_{xx}c_{yy} - c_{xy}^2\right)} \right] \tag{3.9}$$

Then, eccentricity can be computed by equation (3.10).

$$E = \frac{\lambda_2}{\lambda_1} \tag{3.10}$$

Figure 3.4 shows the eccentricity of some shapes.

3.4.2 Minimum Bounding Rectangle (MBR)

MBR is the smallest rectangle that comprises each shape point [7]. For a random shape, eccentricity (E) is the ratio of the length L and width W of the shape's MBR at some group of rotations as shown in equation (3.11).

$$E = \frac{L}{W} \tag{3.11}$$

Elongation (Elo) is another perception created on eccentricity and can be expressed as shown in equation (3.12).

$$\text{Elo} = 1 - \frac{W}{L} \tag{3.12}$$

Elongation values will be in the range of [0, 1]: 0 for symmetrical shapes and closer to 1 for shapes with big aspect ratios.
  Figure 3.5 shows the MBR and corresponding parameter for elongation.

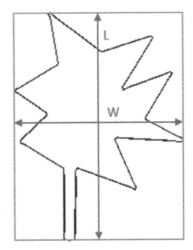

FIGURE 3.5
MBR and corresponding parameter for elongation.

There are some advantages and limitations of this method that are presented here.

Advantages: Eccentricity is robust to noise.

Limitations: This method usually can only discriminate shapes with large variations. This method is not suitable for standalone shape descriptors. Furthermore, the matching cost using this method is high due to the complex normalization of rotation invariance.

3.5 Circularity Ratio (CR)

CR signifies the comparability of the shape to a circle [8]. There are three explanations.

1. CR is the shape area to the circle area ratio having an identical perimeter as represented in equation (3.13).

$$C_1 = \frac{A_s}{A_c} \tag{3.13}$$

where A_s is the shape area and A_c is the circle area having an identical perimeter as the shape. Let the perimeter be P, so $A_c = P^2/4\pi$. Then $C_1 = 4\pi \cdot A_s/P^2$ and 4π is static.

2. CR is the shape area to the shape's perimeter ratio square as shown in equation (3.14).

$$C_2 = \frac{A_s}{P^2} \qquad (3.14)$$

3. CR is also called circle variance and demarcated as shown in equation (3.15).

$$C_3 = \frac{\sigma}{\mu} \qquad (3.15)$$

where σ and μ are the standard deviation and mean of the radial distance from the shape centroid (R_x, R_y) to the contour points (x_i, y_i), $i \in [0, M-1]$. They are represented by equation (3.16).

$$\mu = \frac{1}{M}\sum_{i=0}^{M-1} D_i; \sigma = \sqrt{\frac{1}{M}\sum_{i=0}^{M-1}(D_i - \mu)^2}$$

$$\text{where: } D_i = \sqrt{(x_i - R_x)^2 + (y_i - R_y)^2} \qquad (3.16)$$

Figure 3.6 shows the approach of circularity.

Figure 3.7 shows the circularity amount of some shapes.

There are some advantages and limitations of this method that are presented here.

FIGURE 3.6
Approach of circularity.

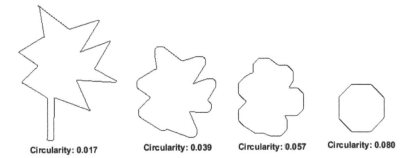

Circularity: 0.017   Circularity: 0.039   Circularity: 0.057   Circularity: 0.080

FIGURE 3.7
Circularity amount of some shapes.

Advantages: This method is easy, straightforward, and invariant to similarity transformation.

Limitations: This method usually can only discriminate shapes with large variations. This method is not suitable for standalone shape descriptors.

3.6 Ellipticity

3.6.1 Ellipse Variance (EV)

EV is a projection error of a shape to fit an ellipse that has an equivalent covariance matrix as the shape $C_{shape} = C$ (equation 3.8) as shown in equation (3.17) [9].

$$EV = \frac{\sigma}{\mu}$$

where:

$$\mu = \frac{1}{M} \sum_{i=0}^{M-1} D'_i$$

$$\sigma = \sqrt{\frac{1}{M} \sum_{i=0}^{M-1} (D'_i - \mu')^2}$$

$$D'_i = \sqrt{L_i^T \cdot C_{shape}^{-1} \cdot L_i}$$

$$L_i = \begin{pmatrix} x_i - R_x \\ y_i - R_y \end{pmatrix}$$

(3.17)

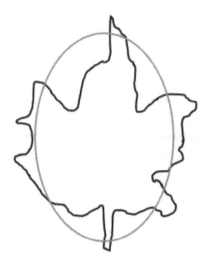

FIGURE 3.8
The ellipticity approach.

EV signifies a shape more accurately than circularity.
    Figure 3.8 shows the ellipticity approach of a shape.

3.6.2 Ellipticity Based on Moment Invariants

Shape ellipticity measure, can be computed from the first two Hu moment invariants. The geometric moments, $m_{p,q}(S)$, of a shape S are defined by equation (3.18) [10].

$$m_{p,q}(S) = \iint\limits_{S} x^p y^q dx dy \tag{3.18}$$

Since any ellipse can be attained by applying an affine transform to a circle, shape ellipticity can be obtained from the affine moment invariant of the circle. Affine invariance can be defined by equation (3.19).

$$I(S) = \frac{\left(m_{2,0}(S).m_{0,2}(S) - m_{1,1}(S)^2\right)}{m_{0,0}(S)^4} \tag{3.19}$$

The ellipticity (E_M) measure can be defined by equation (3.20).

$$E_M(S) = \min\left\{16\pi^2 I(S), \left(16\pi^2 I(S)\right)^{-1}\right\} \tag{3.20}$$

which ranges over [0, 1], peaking at 1 for a perfect ellipse.

There are some advantages and limitations of this method that are presented here.

Advantages: This method is easy, straightforward, and invariant to similarity transformation. This method is more suitable than circularity.

Limitations: This method usually can only discriminate shapes with large variations. This method is not suitable for standalone shape descriptors.

3.7 Rectangularity

Rectangularity denotes how rectangular a shape is.

3.7.1 Smallest Bounding Rectangle (SBR)

SBR signifies how much the shape fills its SBR and can be represented by equation (3.21) [11].

$$SBR = \frac{A_s}{A_{br}} \tag{3.21}$$

where A_s is the shape area and A_{br} is the smallest bounding rectangle area.

A weakness of using the SBR is that it is very delicate to protrusions from the region and sensitive to noise. Even a narrow spike sticking out of a region can massively inflate the area of the SBR and thereby generate very poor rectangularity estimations.

3.7.2 Rectangular Discrepancy Method (RDM)

In an effort to overcome the sensitivity of SBR to noise, RDM is a substitute in which a rectangle is fitted to the region constructed on its moments [12]. Rectangularity is then computed as the normalized discrepancies between the areas of the rectangle and region. More exactly, it is specified in the following areas: D, the difference among the rectangle and the region; R, the difference between the region and the rectangle; and A, the rectangle's area. Then RDM can be expressed by equation (3.22).

$$RDM = 1 - \frac{D+R}{A} \tag{3.22}$$

To overcome variable orientation estimates, the technique was considerably enhanced by considering the original orientation estimation both with and without a 45° offset. The maximum of the two is reserved as the final rectangularity measure is RDM'.

3.7.3 Robust Smallest Bounding Rectangle (RSBR)

An alternative method to overcome the sensitivity of the SBR is to decrease the requirement that the SBR must comprise all the points [13]. If the SBR is required to only encompass the majority of the region, then it should be more robust in the occurrence of small area deviations in the boundary. The formulation of the criterion for the SBR is similar to the previous measure but modified to $(D + R)/I$ where the denominator I is the area of intersection of the region and the rectangle rather than the area of the rectangle. This expression delivers a trade-off between forcing the rectangle to comprise most of the data while keeping the rectangle as small as probable. The RSBR is created by initiating with the standard SBR rescaled to half its area. The concluding fit delivers the rectangularity as represented by equation (3.23).

$$\text{RSBR} = 1 - \frac{D+R}{I} \qquad (3.23)$$

Figure 3.9 shows the rectangularity amount of some shapes.

There are some advantages and limitations of this method that are presented here.

Advantages: This method is easy, straightforward, and invariant to similarity transformation.

Limitations: This method usually can only discriminate shapes with large variations. This method is not suitable for standalone shape descriptors. When the bounding rectangles are used, a high sensitivity to the boundary defect is expected.

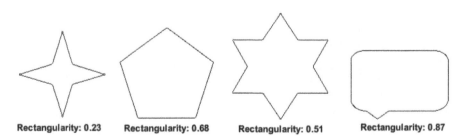

Rectangularity: 0.23   Rectangularity: 0.68   Rectangularity: 0.51   Rectangularity: 0.87

FIGURE 3.9
The rectangularity amount of some shapes.

3.8 Convexity

Convexity is defined as the ratio of convex hull perimeter C_{CH} over that of the original contour C_O as shown in equation (3.24) [14].

$$\text{Convexity} = \frac{C_{CH}}{C_O} \qquad (3.24)$$

The region R is convex only if the whole line segment $P_1 P_2$ is inside the region for any two points $P_1, P_2 \in R$. The region's convex hull is the minimum convex region including it.

Figure 3.10 illustrates the approach of a convex hull.

Figure 3.11 shows the convexity amount of some shapes.

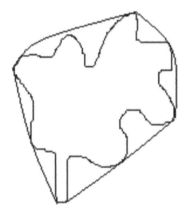

FIGURE 3.10
Illustration of convex hull.

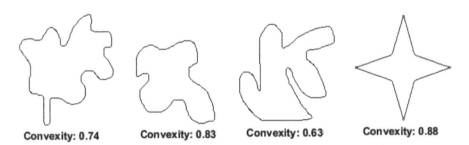

Convexity: 0.74   Convexity: 0.83   Convexity: 0.63   Convexity: 0.88

FIGURE 3.11
The convexity amount of some shapes.

There are some advantages and limitations of this method that are presented here.

Advantages: This method is easy, straightforward, and invariant to geometric transformation.

Limitations: This method usually can only discriminate shapes with large variations. This method is not suitable for standalone shape descriptors.

3.9 Solidity

Solidity defines to what extent the shape is concave or convex, and it is illustrated by equation (3.25) [15].

$$\text{Solidity} = \frac{A_{SR}}{C_{CH}} \tag{3.25}$$

where A_{SR} is the shape region's area and C_{CH} is the shape's convex hull area. The convex shape's solidity is 1 at all times.

Figure 3.12 shows the solidity amount of some shapes.

There are some advantages and limitations of this method that are presented here.

Advantages: This method is easy, straightforward, and invariant to geometric transformation.

Limitations: This method usually can only discriminate shapes with large variations. This method is not suitable for standalone shape descriptors.

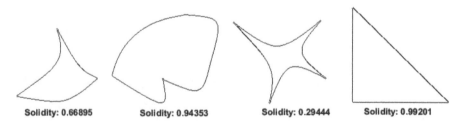

Solidity: 0.66895   Solidity: 0.94353   Solidity: 0.29444   Solidity: 0.99201

FIGURE 3.12
The solidity amount of some shapes.

Euler Number: 1   Euler Number: -1   Euler Number: 0   Euler Number: 0   Euler Number: 1

FIGURE 3.13
Euler number of some shapes.

3.10 Euler Number (EN)

EN defines the relationship among the number of adjoining parts and the number of shape holes [16]. Let P be the number of adjoining parts and H be the number of shape holes. EN is then defined by equation (3.26).

$$\text{Euler} = P - H \tag{3.26}$$

Figure 3.13 shows the Euler number of some shapes.

There are some advantages and limitations of this method that are presented here.

Advantages: This method is easy, straightforward, and invariant to geometric transformation.

Limitations: This method usually can only discriminate shapes with large variations. This method is not suitable for standalone shape descriptors.

3.11 Profiles

The profiles are the shape's projection to the y-axis (vertical projection) and x-axis (horizontal projection) on a Cartesian coordinate system [17]. Let $I(i, j)$ be the area of the shape, then the profile is represented by equation (3.27).

$$\text{Profile}_x(i) = \sum_{j=j_{min}}^{j_{max}} I(i, j)$$

$$\text{Profile}_y(j) = \sum_{i=i_{min}}^{i_{max}} I(i, j) \tag{3.27}$$

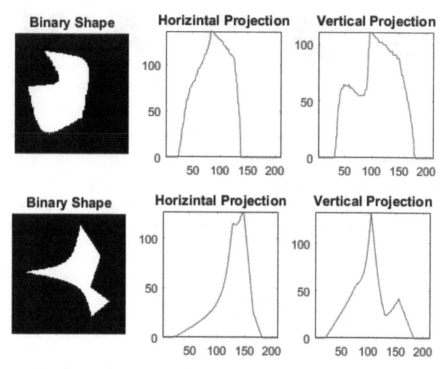

FIGURE 3.14
Profiles of some binary shapes.

Figure 3.14 shows the profiles of some binary shapes.

There are some advantages and limitations of this method that are presented here.

Advantages: This method is easy, straightforward, and invariant to geometric transformation.

Limitations: This method is not suitable for standalone shape descriptors.

3.12 Hole Area Ratio (HAR)

HAR is defined as shown in equation (3.28) [18].

$$HAR = \frac{A_H}{A_S} \qquad (3.28)$$

9	0	B	Q
Hole-Area Ratio: 0.10	Hole-Area Ratio: 0.26	Hole-Area Ratio: 0.12	Hole-Area Ratio: 0.22

FIGURE 3.15
Hole area ratio of some shapes.

where A_S is the shape's area and A_H is the entire area of all shape holes. HAR is most efficient in differentiating among symbols with large holes and symbols that have small holes.

Figure 3.15 shows the hole area ratio of some shapes.

There are some advantages and limitations of this method that are presented here.

Advantages: This method is easy, straightforward, and invariant to geometric transformation.

Limitations: This method usually can only discriminate shapes with large variations. This method is not suitable for standalone shape descriptors.

3.13 Summary

This chapter presents a technique for distinguishing an object by using some geometrical shape features. Generally, the simple geometric descriptor can only differentiate shapes with big variances; thus, they are generally utilized as filters to remove false hits or to differentiate shapes united with other shape descriptors. They aren't appropriate to be standalone shape descriptors. A shape can be designated by various features. These shape parameters are CoG, AMI, ABE, Eccentricity, GR, EV, Rectangularity, Convexity, Solidity, EN, Profiles, and HAR.

References

1. Araki, T., Ikeda, N., Molinari, F., Dey, N., Acharjee, S. M., Saba, L., Kumar, D. et al. (2014). Effect of geometric-based coronary calcium volume as a feature along with its shape-based attributes for cardiological risk prediction from low contrast intravascular ultrasound. *Journal of Medical Imaging and Health Informatics*, 4(2), 255–261.

2. Chaki, J., Parekh, R., & Bhattacharya, S. (2015). Plant leaf recognition using texture and shape features with neural classifiers. *Pattern Recognition Letters, 58,* 61–68.

3. Altun, O., Albayrak, S., Ekinci, A., & Bükün, B. (2006). Turkish fingerspelling recognition system using axis of least inertia based fast alignment. In *Australasian Joint Conference on Artificial Intelligence* (pp. 473–481). Springer, Berlin, Germany.

4. Tian, Z., Dey, N., Ashour, A. S., McCauley, P., & Shi, F. (2018). Morphological segmenting and neighborhood pixel-based locality preserving projection on brain fMRI dataset for semantic feature extraction: An affective computing study. *Neural Computing and Applications, 30*(12), 3733–3748.

5. Bagal, V. C., & Manza, R. R. (2014). A survey on plant species classification based on morphological features of leaf. *International Journal of Advances in Engineering & Technology, 7*(2), 553.

6. Jiji, G. W., & DuraiRaj, P. J. (2015). Content-based image retrieval techniques for the analysis of dermatological lesions using particle swarm optimization technique. *Applied Soft Computing, 30,* 650–662.

7. Saba, L., Dey, N., Ashour, A. S., Samanta, S., Nath, S. S., Chakraborty, S., Sanches, J., Kumar, D., Marinho, R., & Suri, J. S. (2016). Automated stratification of liver disease in ultrasound: An online accurate feature classification paradigm. *Computer Methods and Programs in Biomedicine, 130,* 118–134.

8. Kopanja, L., Žunić, D., Lončar, B., Gyergyek, S., & Tadić, M. (2016). Quantifying shapes of nanoparticles using modified circularity and ellipticity measures. *Measurement, 92,* 252–263.

9. Huang, F., Gan, Y., Zhang, D., Deng, F., & Peng, J. (2018). Leaf shape variation and its correlation to phenotypic traits of soybean in Northeast China. In *Proceedings of the 2018 6th International Conference on Bioinformatics and Computational Biology* (pp. 40–45). ACM, New York.

10. Žunić, J., Kakarala, R., & Aktaş, M. A. (2017). Notes on shape based tools for treating the objects ellipticity issues. *Pattern Recognition, 69,* 141–149.

11. Schmith, J., Höskuldsson, Á., & Holm, P. M. (2017). Grain shape of basaltic ash populations: Implications for fragmentation. *Bulletin of Volcanology, 79*(2), 14.

12. Xiao, Q., Pan, J., Lv, Z., Xu, J., & Wang, H. (2017). Measure of bubble non-uniformity within circular region in a direct-contact heat exchanger. *International Journal of Heat and Mass Transfer, 110,* 257–261.

13. Barequet, G., & Har-Peled, S. (2001). Efficiently approximating the minimum-volume bounding box of a point set in three dimensions. *Journal of Algorithms, 38*(1), 91–109.

14. Kumar, R., Talukdar, F. A., Dey, N., & Balas, V. E. (2016). Quality factor optimization of spiral inductor using firefly algorithm and its application in amplifier. *International Journal of Advanced Intelligence Paradigms,* 1–17.

15. El Naqa, I., Grigsby, P. W., Apte, A., Kidd, E., Donnelly, E., Khullar, D., Chaudhari, S. et al. (2009). Exploring feature-based approaches in PET images for predicting cancer treatment outcomes. *Pattern Recognition, 42*(6), 1162–1171.

16. Maji, P., Chatterjee, S., Chakraborty, S., Kausar, N., Samanta, S., & Dey, N. (2015). Effect of Euler number as a feature in gender recognition system from offline handwritten signature using neural networks. In *2015 2nd International Conference on Computing for Sustainable Global Development (INDIACom)* (pp. 1869–1873). IEEE.

17. Morin, A. J., & Marsh, H. W. (2015). Disentangling shape from level effects in person-centered analyses: An illustration based on university teachers' multidimensional profiles of effectiveness. *Structural Equation Modeling: A Multidisciplinary Journal, 22*(1), 39–59.

18. Khestanova, E., Guinea, F., Fumagalli, L., Geim, A. K., & Grigorieva, I. V. (2016). Universal shape and pressure inside bubbles appearing in van der Waals heterostructures. *Nature Communications, 7,* 12587.

4

Polygonal Approximation Shape Features

Polygonal approximation (PA) can be used to overlook the insignificant edge discrepancies, and as a substitute capture the complete shape. This is beneficial as it decreases the distinct pixelization effect of the outline of the shape or contour [1]. PA methods can also be utilized as pre-processing techniques for further shape feature extraction.

An open M-vertex polygonal curve A in 2-D planetary is denoted as the well-ordered group of vertices $A = \{a_1, ..., a_M\} = \{(x_1, y_1), ..., (x_M, y_M)\}$. The result of polygonal approximation is a curve B comprised of N vertices: $B = \{b_1, ..., b_N\}$, where the group of vertices b_N is a subgroup of A and $N < M$. The end points of B are the end points of A, i.e., $b_1 = a_1$ and $b_N = a_M$.

PA methods can be utilized as a simple technique for edge explanation and representation. There are two categories of optimization problems associated with polygonal approximation problems:

Min-ε problem: Given a polygonal curve A, approximate it by another polygonal curve B with a specified number of line segments S so that the approximation error $E(A)$ is reduced.

Min-# problem: Given a polygonal curve A, approximate it by another polygonal curve B with the least number of segments S so that the approximation error $E(A)$ doesn't exceed a specified maximum tolerance ε.

There are two procedures to estimate the excellence of polygonal approximation algorithms. (1) *Fidelity* calculates how well the approximated polygon fits the curve comparative to the optimal polygon in terms of the approximation error. (2) *Efficiency* calculates how dense the approximated polygonal presentation of the curve is. They are demarcated as shown in equation (4.1):

$$\text{Fidelity} = \frac{E_{min}}{E} \times 100$$

$$\text{Efficiency} = \frac{S_{min}}{S} \times 100$$

(4.1)

where S is the number of sections for an algorithm under question, S_{min} is the number of segments for min-# solution, E is the approximation error for an approximation algorithm, and E_{min} is the approximation error of the optimal solution for the min-ε problem.

Some polygonal approximation methods are discussed below.

4.1 Merging Method (MM)

Merging approaches can form a line segment by joining consecutive pixels if every new pixel which is appended does not cause the section to diverge from the straight line too much [2]. Merging techniques generally adhere to the following steps:

Step 1: Merge points along an edge until the least square error line acceptable to the points merged so far exceeds a threshold.

Step 2: Record the two end points of the line.

Step 3: Repeat Steps 1 and 2 until all boundary points are handled.

Figure 4.1 demonstrates the merging polygonal approximation technique steps. The line represents the approximated polygon.

Some merging polygonal approximation techniques are discussed below.

4.1.1 Distance Threshold Method (DTM)

DTM first selects one point as an initial point on the contour of the shape [3]. Then, for every new point which is appended with the initial point, a line is drawn to this new point from the initial point. After that, for each point along the line/segment the squared error is computed. If the error is greater than some predefined threshold, the line from the initial point to the preceding point is kept and a new line is started.

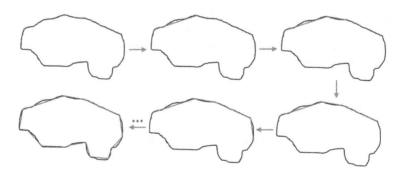

FIGURE 4.1
Consecutive steps of merging polygonal approximation technique, starting from the top left image and ending at the bottom left image.

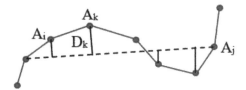

FIGURE 4.2
Diagram of the distance from a point on the border to a linear segment.

The maximum applied error measures in usage are built on distance among the approximated linear segments and the original input curve vertices. The distance D_k (i, j) from curve vertex $A_k = (x_k, y_k)$ to the equivalent approximation linear segments (A_i, A_j) is defined as shown in equation (4.2). A diagram of the distance to a linear segment from a point on the border is shown in Figure 4.2.

$$D_k(i,j) = \frac{\left|(x_j - x_i)(y_i - y_k) - (x_i - x_k)(y_j - y_i)\right|}{\sqrt{(x_j - x_i)^2 + (y_j - y_i)^2}} \tag{4.2}$$

4.1.2 Tunnelling Method (TM)

TM is applicable if the border of a shape is thick enough instead of single-pixel border [4]. The concept is the same as laying the least number of straight rods end to end along a curved tunnel. We select one point as the starting point and lay as elongated a straight rod as possible. For the tunnel curvature, it is not possible to cover the entire tunnel with a single rod, so another rod is laid and another until the end is reached.

Both the tunnelling method and distance threshold method can accomplish polygonal approximation proficiently.

4.1.3 Polygon Evolution by Vertex Deletion (PEVD)

The main idea of PEVD is quite simple: in each evolution phase, a couple of successive line segments (the line between two successive vertices is the line segment) L_1 and L_2 are replaced by a single line segment merging the endpoints of L_1 and L_2. The significance of this method is the order of the replacement [5]. The replacement is finished conferring to a significance measure M given by equation (4.3).

$$M(L_1, L_2) = \frac{\beta(L_1, L_2)S(L_1)S(L_2)}{S(L_1) + S(L_2)} \tag{4.3}$$

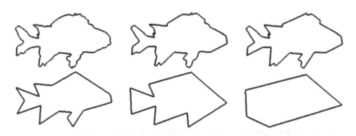

FIGURE 4.3
Steps of polygon evolution, starting from top left ending at bottom right.

where $\beta(L_1, L_2)$ is the turn angle at the mutual segments' vertexes L_1, L_2 and $S(L_1)$, $S(L_2)$ is the length of L_1 or L_2, normalized with regard to the entire polygonal curve length. The evolution procedure presumes that vertices that are enclosed by segments with a large value of $M(L_1, L_2)$ are vital if they are not of a low value. Figure 4.3 demonstrates an example.

The polygon evolution [6] technique accomplishes the shape generalization task, i.e., the evolution procedure equates the importance of the contour vertices constructed on a significance measure. As any digital curve can be represented as a polygon without information loss (with probably a bulky number of vertices), it is enough to examine the polygonal shapes' evolutions for shape feature extraction.

There are some advantages and limitations of these methods that are presented here.

Advantages: They represent the shape with less complexity and no blurring effects. They eliminate noise during approximation. There is no displacement of appropriate or relevant features, even though inappropriate or irrelevant features disappear after polygonal approximation. After polygonal approximation, the residual vertices on an edge don't alter their locations.

Limitations: The biggest shortcoming is that the vertices do not always correspond to actual corners in the boundary.

4.2 Splitting Method (SM)

SM is an iterative procedure that divides the curve into smaller and smaller curves repeatedly until the maximum perpendicular distances of the points on the curve from the line segment are smaller than the error tolerance. The procedure is repeated recursively for each of the two new lines until no break is needed. Figure 4.4 demonstrates an instance. Figure 4.5 shows the tree structure for Figure 4.4.

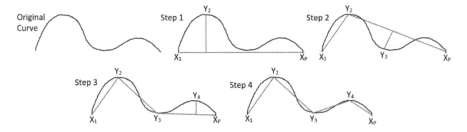

FIGURE 4.4
Splitting method for polygonal approximation. In step 1 Y_2 is the farthest point from the segment and the same for steps 2, 3, and 4.

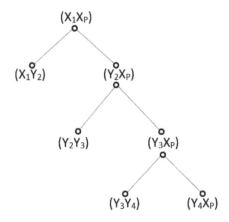

FIGURE 4.5
Tree structure for Figure 4.4.

Figure 4.6 shows the splitting technique of a closed curve. First the farthest points of the shape are found and joined by a line (X_1X_P). Then from each side of X_1X_P, the farthest point is recorded $(Y_1$ and $Z_1)$. The points are connected by line segments: X_1Z_1, Z_1X_P, X_PY_1, $Y_{1 \times 1}$. For every segment, the distance between its points is checked. If the corresponding points from the original border are smaller than a threshold, then stop; otherwise, subdivide the segment further.

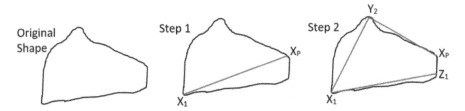

FIGURE 4.6
The splitting technique of a closed curve: Y_1 and Z_1 are the farthest points from X_1X_P.

There are some advantages and limitations of this method that are presented here.

> *Advantages*: The algorithm works in any dimension as it only depends on computing the distance between points and lines.

> *Limitations*: The main limitation of this method is the dependency on the starting point. It also suffers from stressing outliers.

4.3 Minimum Perimeter Polygon (MPP)

MPP approximation offers good depictions for image analysis applications [7]. The idea is to enclose a boundary by a group of concatenating cells. The boundary is permitted to shrink, but it is controlled by the outer and inner walls of the bounding area demarcated by the cells. Finally, the shape shrinking reveals the MPP. The vertices of the MPP overlap with the corners of the outer or the inner wall. The size of the cells controls the accuracy of the depiction. The aim is to utilize the highest possible cell size suitable for a specified application. The shape of the object surrounded by the inner wall of the light gray cells is represented in dark gray (see Figure 4.7). Navigating the boundary in counter clockwise direction, we come to the convex (represented with white dots) or concave (represented with black dots) vertices. The vertices of the MPP overlap either with convex vertices in the inner wall or with the mirrors of the concave vertices in the outer wall.

The orientation of triplets of points is represented by equation (4.4).

$$p = (x_1, y_1), q = (x_2, y_2), r = (x_3, y_3)$$

$$S = \begin{bmatrix} x_1 & y_1 & 1 \\ x_2 & y_2 & 1 \\ x_3 & y_3 & 1 \end{bmatrix} \tag{4.4}$$

$$det(S) = \begin{cases} > 0 & (p,q,r) \text{ is a counterclockwise direction} \\ = 0 & (p,q,r) \text{ are colinear} \\ < 0 & (p,q,r) \text{ is a clockwise direction} \end{cases}$$

Here $det(S)$ is represented as $sgn(p, q, r)$. $sgn(p, q, r) > 0$ designates that point r lies on the positive side of the line passing from (p, q). $sgn(p, q, r) < 0$ designates that point r lies on the negative side of the line passing from (p, q).

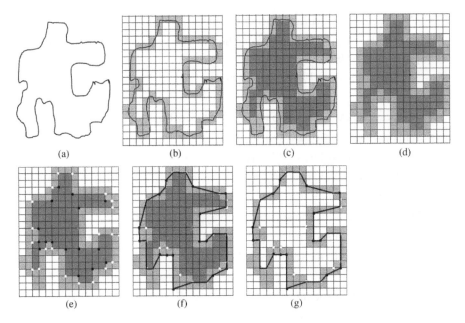

FIGURE 4.7
(a) An object boundary (black curve); (b) Boundary enclosed by cells (in gray); (c, d) Region (dark gray) resulting from enclosing the original boundary by cells; (e) Convex (white dots) and concave (black dots) vertices obtained by following the boundary of the dark gray region in the counter clockwise direction; (f) Concave vertices (black dots) displaced to their diagonal mirror location in the outer wall of the boundary region; convex vertices are not changed; (g) The result of polygonal approximation using MPP.

4.3.1 Data Preparation for MPP

- List of coordinates of every vertex.
- Label every vertex as W (convex) or B (concave).
- List of the mirrors of B vertices.
- Vertices essentially are in sequential order.
- The first vertex V_0 is the uppermost, leftmost vertex.
- The algorithm utilizes a black crawler B_C and a white crawler W_C crawling along the mirrored concave (B) and convex (W) vertices respectively.

4.3.2 MPP Algorithm

- Initialization: $W_C = B_C = V_0$.
- V_L is the final vertex observed.

- V_K is the present vertex being observed.
- $sgn(V_L, W_C, V_K) > 0$ denotes V_K lies to the positive side of the line passing through (V_L, W_C).
  - The subsequent MPP vertex is W_C.
  - $V_L = W_C$.
  - Continue with the subsequent vertex after V_L.
- $sgn(V_L, W_C, V_K) \leq 0$ and $sgn(V_L, B_C, V_K) \geq 0$.
  - V_K is a candidate MPP vertex.
  - If V_K is convex, then $W_C = V_K$.
  - Otherwise $B_C = V_K$.
  - Continue with the subsequent vertex in the list.
- $sgn(V_L, W_C, V_K) \leq 0$ and $sgn(V_L, B_C, V_K) < 0$.
  - B_C turns into a candidate MPP vertex.
  - $V_L = B_C$.
  - Reinitialize the algorithm by setting $W_C = B_C = V_L$.
  - Continue with the next vertex in the list.
- Continue until the initial vertex is reached again.

There are several advantages and limitations of this method as described as follows:

Advantages: It represents the shape with less complexity.

Limitations: This is a time consuming approach. It is difficult to set the proper size of the cells, which determines the accuracy of the representation.

4.4 Dominant Point (DP) Detection

In DP detection, maximum shape information is confined in the corners (high curvature points), which are able to characterize the border or contour of the shape [8]. To approximate curves with straight lines, high curvature points are the best place at which to break the lines. There are two key steps in these algorithms to attain the dominant points. One is to calculate the measure of relative consequence (e.g., curvature), and the other is to attain the region of support for calculating the measure of relative consequence. To calculate curvature, one must describe the region of support. The dominant points attained from the vertices represents the polygon approximation.

There are some advantages and limitations of this method that are presented here.

Advantages: In a time sequence, dominant points can be used to compute the displacement between each pair of consecutive images.

Limitations: The dominant point algorithms have the limitation that they spot either less or more points than are really present. The reason for this is that curvature is a local property and is delicate to local variations. A small region of support can't take care of local discrepancies and will consequently generate more dominant points. A large region of support will cause loss of accuracy of localization so that dominant points matching to fine features cannot be found, thus generating fewer dominant points than are really present. Additionally, the algorithm is not integrally parallel in nature.

4.5 K-means Method

The K-means based clustering technique is used to divide the contour into P subgroups of points such that every subgroup can be fitted by a straight line [9]. First, the points are divided into P random clusters by selecting extreme curvatures as the break points (see Figure 4.8). Then the line fitting approaches are developed to solve one of the following issues. (1) Type 1 issue: for a specified tolerance of error, the aim is to reduce the number of approximating line segments. (2) Type 2 issue: for a specified number of approximating line segments, the aim is to reduce the error norm among the input digital curve and the approximating polygon. (3) Type 3 issue: approximate the input digital curve according to its own qualities without any predefined constraint. Type 3 issue is more vital in real applications as the required number of approximating line segments and the approximation error are not easy to determine prior to seeing the results. The solutions of all three issues can be supposed as an optimization procedure of finding for appropriate vertices of a polygon from the input

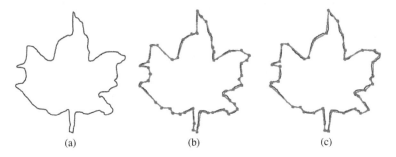

(a)   (b)   (c)

FIGURE 4.8
Polygonal approximation for the shape. (a) Original image; (b) #DP = 52; (c) #DP = 39.

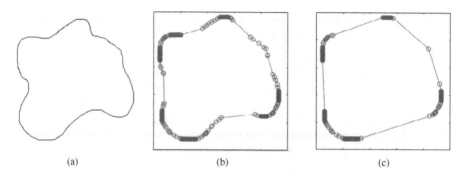

(a)   (b)   (c)

FIGURE 4.9
Polygon approximation by K-means. (a) Original shape boundary; (b) Approximation amount = 0.6; (c) Approximation amount = 0.1.

digital curve. A complete search for the optimal solution in the parameter space will show an exponential complexity; thus, most of the present polygonal approximation approaches yield suboptimal outcomes to save computational cost (Figure 4.9).

There are some advantages and drawbacks of this method, which is presented as follows:

Advantages: If the value of k is small, then K-means is most often the computationally faster method.

Limitations: It is difficult to predict the value of K.

4.6 Genetic Algorithm (GA)

Suppose a curve $A = \{a_1, a_2, ..., a_i, ..., a_j, ..., a_m\}$ is comprised of m points well-arranged along the curve in a clockwise direction [10]. Consider that $a_i a_j$ and $A_i A_j$ denote the chord and arc initiating from a_i and to a_j along the curve path in a clockwise direction. The purpose of this method is to detect an approximate polygon P^n containing n breakpoints from the specified curve A for the specified object curve and the number of breakpoints n, so that the entire arc-to-chord deviation among the curve and the polygon is reduced.

The following components should be found in a GA for a specific problem:

1. A technique for encoding the solution as a binary string
2. Fitness function
3. Genetic operators or control parameters

4.6.1 Encoding

The encoding technique is utilized to map every approximated polygon into a unique binary string [11]. Here, a binary string is demarcated as a polygon chromosome. An encoding technique should be outlined in such a way that it is effective, reasonable, and competent for the process of optimization. The entire chromosome length is the number of original curve points m. Every bit of the chromosome denotes a curve point. In the binary string, the value 1 of a bit designates that the curve point is the approximated polygon breakpoint. The chromosome, the sum of the bit values that is n, signifies a polygon with n line segments. For example, the chromosome in Figure 4.10 (right) is 101011001101.

The arc-to-chord divergence from point a_i to point a_j, represented as D_{ij} is denoted as shown in equation (4.5).

$$D_{ij} = \sum_{a_k} (S_k)^2, \forall a_k \in A_i A_j \tag{4.5}$$

where S_k represents the orthogonal distance of a point a_k on the original curve to its equivalent chord $a_i a_j$.

The entire arc-to-chord deviation among approximated polygon P^n and original curve A is represented by the integral square error as shown in equation (4.6).

$$E = \sum_{i=1}^{m} (S_i)^2 \tag{4.6}$$

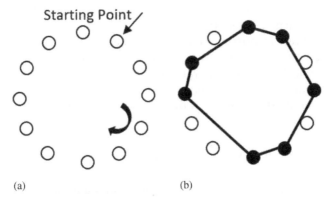

(a)   (b)

FIGURE 4.10
(a) Original curve ($m = 12$); (b) Approximated polygon ($n = 7$).

4.6.2 Fitness

The fitness function $F(C)$ of every chromosome C calculated on g-th generation is denoted as shown in equation (4.7).

$$F(C) = Max(E(g)) - E(C) \tag{4.7}$$

where $E(C)$ is the integral square error of the equivalent chromosome polygon C. $Max(E(g))$ is the highest integral square error among all g-th generation chromosomes.

4.6.3 Genetic Operators or Control Parameters

Genetic operators or control parameters include selection, crossover, and mutation. A roulette wheel is used for the selection procedure, where every chromosome has a slot that is proportional to its fitness. The probability that chromosome C_i ($i = 1,2,... M$, M = maximum size) is carefully chosen as a next generation member is defined as shown in equation (4.8).

$$R_i = \frac{F(C_i)}{\sum\limits_{j=1}^{M} F(C_j)} \tag{4.8}$$

The M chromosomes are carefully chosen and then pass to a mating pool for additional genetic operations. The successor step is the crossover operation after the selection process. To continue the crossover operation, two steps are needed. First of all, couples of chromosomes are selected arbitrarily in the mating pool. Then, a crossover point is arbitrarily chosen for every pair of chromosomes between 1 to the chromosome length minus 1. The crossover operation data exchange delivers a great deal of the power in genetic algorithms. In the optimization process, crossover has an important role. Lastly, the mutation process includes altering the values of genes in a chromosome: it implies altering a 0 to 1 and vice versa. These three parameters of the genetic algorithm need to be adjusted carefully for efficiency.

There are some advantages and limitations of this method, which are presented as follows:

Advantages: This method performs a parallel search in complex parameter space.

Limitations: The selection, crossover, and mutation parameters of the genetic algorithm need to be adjusted carefully for efficiency, which is a challenging job.

4.7 Ant Colony Optimization (ACO) Method

The study of actual ant behavior informs the ACO method [12]. Ethologists detected that ants can build the shortest way from their position to the source of the food and back by the use of marks. An ant leaves marks on the ground and detects the way by this trail. The next ant will recognize the mark placed on various paths and selects one with a probability proportionate to the amount of marks on it. Then the ant passes through the selected path and leaves a mark itself. This is a recursive procedure which recognizes the path along which the most ants formerly passed through. The overall principles for the ACO simulation method of actual ant behavior are described below.

4.7.1 Initialization

There are two parts to initialize the ACO method: the problem graph depiction and the preliminary ant circulation. First, the fundamental problem is denoted by a graph, $G = (N, E)$, where N and E represent the node groups and the edge groups respectively. The graph is linked, but may or may not be complete, so that the possible answers to the original problem resemble to graph paths that favor the problem-domain constraints. Second, several ants are randomly positioned on the nodes selected arbitrarily. Then the distributed ants will accomplish a graph tour by creating a path conferring to the node transition rule defined next.

4.7.2 Node Transition Rule

The ants change their position from one node to another grounded using this rule. According to the problem-domain constraints, few nodes could be demarked as unreachable for a walking ant. For instance, the nodes that an ant has visited are marked as unreachable to permit the ant to go to each node of the graph in limited steps. The transition rule for nodes is probabilistic. For the i-th ant on node k, the choice of the next node n to trail depends on the probability of the node transition, which can be represented by equation (4.9).

$$P_{kn}^i = \begin{cases} \dfrac{\left(\tau_{kn}\right)^{\alpha}\left(\eta_{kn}\right)^{\beta}}{\displaystyle\sum_{s \notin tabu_i}\left(\tau_{ks}\right)^{\alpha}\left(\eta_{ks}\right)^{\beta}}, & \text{if } n \notin tabu_i \\[4pt] 0, & \text{otherwise} \end{cases}$$

(4.9)

where τ_{kn} is the mark strength of on the edge (k, n), η_{kn} is the visibility value of the edge (k, n), α and β are controller constraints, and $tabu_k$ denotes the group of presently unreachable nodes for the i-th ant as per the problem-domain constraints. The strength of marks placed on the edge (k, n) denoting the earlier ant knows about this edge is a shared memory that delivers indirect communication among the ants; i.e., by detecting the strength of marks placed on edges, the ants communicate with one another. The visibility value is calculated by a greedy heuristic for the fundamental problem that deliberates only the local information on edge (k, n) such as the length of it. Parameters α and β control the relative influence among the two above-mentioned measurement types.

4.7.3 Pheromone Updating Rule

The ant moves across borders to various nodes by using the node transition rule iteratively until an answer to the fundamental problem is created [13]. For instance, a solution is attained when an ant passes each node during its tour. When each ant creates a solution, an ACO algorithm cycle is finished. The strength of marks on every edge is updated by the mark updating rule defined by equation (4.10) at the end of every cycle.

$$\tau_{kn} \leftarrow \rho\tau_{kn} + \sum_{i=1}^{m} \Delta\tau_{kn}^{i} \qquad (4.10)$$

where $\rho \in (0, 1)$ is the perseverance rate of earlier marks, $\Delta\tau_{kn}^{i}$ is the amount of marks placed on edge (k, n) by the i-th ant at the present cycle, and m is the number of scattered ants. In a real ant system, shorter paths will hold more amounts of marks; thus, in the ACO method, the paths conforming to fitter solutions should obtain more marks and become more attractive in the subsequent cycle.

4.7.4 Stopping Criterion

The stopping criterion of the ACO algorithm could be the maximum number of successive cycles, the CPU time limit, or the highest number of cycles among two enhancements of the best global solution.

The following are some advantages and drawbacks of this method.

Advantages: This method has inherent parallelism. Positive feedback accounts for rapid discovery of good solutions. It is efficient for the traveling salesman problem and similar problems. This method can be utilized in dynamic applications (as it adapts to changes such as new distances, etc.).

Limitations: Theoretical analysis is difficult. Sequences of random decisions are generated. Probability distribution changes by iteration. Research is experimental rather than theoretical. Time to convergence is uncertain (but convergence is guaranteed).

4.8 Tabu Search (TS)

TS initiates from a preliminary solution selected arbitrarily or attained from another polygonal approximation technique [14]. It travels iteratively from one solution to another on the basis of the current solution until a specific stop criterion is met. There is a group of candidate movements, called the neighborhood, to choose from during each iteration of movements. Those candidates who immediately return to the solutions currently visited are referred to as active tabu and are made unattainable by not including them in the neighborhood. This is followed by a specially designed memory mechanism called the tab list (TL). When a move is selected in a single iteration, the route of this move is recorded in the TL, and in the succeeding iterations it will not be reversed. The TL has a static size. When the TL is filled, the first element should be eliminated before the new element can be placed in, i.e., the TL is circular. Therefore, every tabu move will be released and resume accessibility after certain iterations reliant on the TL length. Particularly, the tabu move status can be overruled and made reachable right away if a specific aspiration condition is encountered, such as the tabu move will result in an improved objective value over the best one attained until that date. The steps of the TS algorithm for the polygonal approximation problem are detailed below.

Suppose there are m points in the input digital curve designated as $D = \{x_0, x_1,...,x_{m-1}\}$. The purpose is to discover the minimum subgroup of D as the polygon vertices so that the error norm between D and the polygon is less than a predefined threshold value ε (called the ε-bound constraint). The error norm (ε_{norm}) can be calculated in various techniques, such as maximal perpendicular distance error, integral square error, and the area deviation error.

Every possible solution of the polygonal approximation problem can be signified by a binary string as shown in equation (4.11).

$$\beta = b_0 b_1 \cdots b_{m-1} \tag{4.11}$$

where $b_i = 0$ or 1, and the group of vertices of the polygon is represented by $V = \{x_i | b_i = 1\}$. Therefore, every bit of β is a pointer defining whether the equivalent input point is a vertex. The number of the polygon vertices is equal to the summation of the complete bit values. The approximation error can be calculated by evaluating the value of ε_{norm} among the digital curve and the approximating polygon described by string β.

The construction of the objective function to the polygonal approximation problem is shown using equation (4.12).

$$\text{Mimimize}: f(\beta) = \sum_{i=0}^{m-1} b_i$$

$$\text{Subject\_to}: \varepsilon_{norm} < \varepsilon \tag{4.12}$$

The optimal polygon is defined by the solution that has the minimum value of the objective function ($f(\beta)$) and satisfies the ε-bound constraint [15].

4.8.1 Initialization

TS initiates from a preliminary solution. There are two ways of creating the preliminary solution. One is created on randomization, i.e., arbitrarily choice of a value for every parameter of the solution. The other is created on any local search polygonal approximation result.

4.8.2 Definition of Moves

The move operations are vital for the performance of TS. Precisely designed moves are essential for the application. Three sorts of moves are discussed here. The first and the second types of moves accomplish vertex-deletion and vertex-addition processes to build a search path. The third type of moves granularly modifies the locations of vertices to satisfy the ε-bound constraint. The details are described as follows:

1. A^--*move (vertex deletion process):*

   For every element in V, it is moved to $D{-}V$ related with a vertex-deletion probability P. The value of the objective function will be satisfied by executing this move as the number of vertices is reduced. The value of P controls the reduction rate of the number of vertices. The higher the value of P, the quicker the number of vertices reduces. However, the ultimate objective value will not be good enough if P is too high.

2. A^+-*move (vertex addition process):*

   Arbitrarily choose an element from $D{-}V$, and move it to V. As an alternative to including a probability as used in A^--move, add only one vertex for every A^+-move and keep the number of vertices as small as probable. The value of the objective function will rise by one unit and become inferior by executing this move.

3. B-*move (vertex adjustment process):*

   Arbitrarily move an element from V to $D{-}V$, and vice versa. B-move is an advanced level process which combines A^--move and A^+-move in a sole move. The value of the objective function will not be altered by applying B-move as the number of vertices remains identical. However, the approximation error (ε_{norm}) can be attuned as excellently as required by operating B-moves iteratively.

To make the finest move in every iteration, a group of trials of the selected type of moves is created for the present solution. First, these trial moves are arranged based on their objective values (for A^--move) or approximation errors (for A^+-move and B-move) in increasing order. Therefore, the finest move is the first trial that is not tabu, or which is tabu but satisfies the aspiration criteria described beneath.

4.8.3 Aspiration Criteria (AC)

AC of A^--move: The tabu status of an A^--move is overruled if it results in an improved objective value than the finest one attained so far.

AC of A^+-move: The tabu status of an A^+-move is overruled if it results in an approximation error less than ε.

AC of B-move: The tabu status of a B-move is overruled if it results in an approximation error less than ε.

After the finest move is selected, the route of this move is documented in the bottom of the TL (remove the first element if the TL is full) and should not be reversed.

The following are some advantages and limitations of TS.

Advantages: This method allows the designer to select the maximum number of cells as well as machines in a cell.

Limitations: This method searches all moves that penalize the current implementation when the number of variables is very large.

4.9 Summary

The existing polygonal approximation methods can be separated into two optimal classes:

1. *Classical algorithms*: merge, split, minimum perimeter polygon, and dominant points detection.
2. *Optimization algorithms*: K-means, genetic algorithms, ant colony optimization methods, and tabu search.

The classical algorithms are generally created on experimental approaches or methods. The reliability of these algorithms is generally low as the global optimization error is not measured through the course of the approximation. In optimization algorithms the approximation problem is measured as an optimization task where the global approximation error is the key condition to be controlled. The solution that delivers minimum approximation error can be attained by stochastic optimization approaches (such as ant colony method and genetic algorithms) or by local optimization approaches (such as tabu search). The search can be initiating the search can be attained with some experimental algorithm for approximation, or any arbitrary approximation can be used. Then the preliminary approximation (or approximations) is enhanced to search the smallest global approximation error. Algorithms of this class can deliver near-optimal or occasionally optimal results, but the global optimality can't be definite even in the case of iterative methods.

References

1. Liu, Z., Watson, J., & Allen, A. (2016). A polygonal approximation of shape boundaries of marine plankton based-on genetic algorithms. *Journal of Visual Communication and Image Representation, 41,* 305–313.
2. Chaki, J., & Dey, N. (2018). *A Beginner's Guide to Image Preprocessing Techniques.* CRC Press, Boca Raton, FL.
3. Araki, T., Ikeda, N., Dey, N., Chakraborty, S., Saba, L., Kumar, D., Godia, E.C. et al. (2015). A comparative approach of four different image registration techniques for quantitative assessment of coronary artery calcium lesions using intravascular ultrasound. *Computer Methods and Programs in Biomedicine, 118*(2), 158–172.
4. Shi, D., Gunn, S. R., & Damper, R. I. (2003). Handwritten Chinese radical recognition using nonlinear active shape models. *IEEE Transactions on Pattern Analysis and Machine Intelligence, 25*(2), 277–280.
5. Kamal, M. S., Chowdhury, L., Khan, M. I., Ashour, A. S., Tavares, J. M. R., & Dey, N. (2017). Hidden Markov model and Chapman Kolmogrov for protein structures prediction from images. *Computational Biology and Chemistry, 68,* 231–244.
6. Kamal, S., Ripon, S. H., Dey, N., Ashour, A. S., & Santhi, V. (2016). A MapReduce approach to diminish imbalance parameters for big deoxyribonucleic acid dataset. *Computer Methods and Programs in Biomedicine, 131,* 191–206.
7. Kalyoncu, C., & Toygar, Ö. (2015). Geometric leaf classification. *Computer Vision and Image Understanding, 133,* 102–109.
8. Chaki, J., Parekh, R., & Bhattacharya, S. (2015). Plant leaf recognition using texture and shape features with neural classifiers. *Pattern Recognition Letters, 58,* 61–68.
9. Bose, S., Mukherjee, A., Chakraborty, S., Samanta, S., & Dey, N. (2013, December). Parallel image segmentation using multi-threading and K-means algorithm. In *2013 IEEE International Conference on Computational Intelligence and Computing Research (ICCIC)* (pp. 1–5). IEEE.
10. Dey, N., Ashour, A. S., Beagum, S., Pistola, D. S., Gospodinov, M., Gospodinova, E. P., & Tavares, J. M. R. (2015). Parameter optimization for local polynomial approximation based intersection confidence interval filter using genetic algorithm: An application for brain MRI image de-noising. *Journal of Imaging, 1*(1), 60–84.

11. Karaa, W. B. A., Ashour, A. S., Sassi, D. B., Roy, P., Kausar, N., & Dey, N. (2016). Medline text mining: An enhancement genetic algorithm based approach for document clustering. In *Applications of Intelligent Optimization in Biology and Medicine* (pp. 267–287). Springer, Cham, Switzerland.
12. Lu, D. S., & Chen, C. C. (2008). Edge detection improvement by ant colony optimization. *Pattern Recognition Letters, 29*(4), 416–425.
13. Li, X., Zhu, W., Ji, B., Liu, B., & Ma, C. (2010). Shape feature selection and weed recognition based on image processing and ant colony optimization. *Transactions of the Chinese Society of Agricultural Engineering, 26*(10), 178–182.
14. Li, Z., Dey, N., Ashour, A. S., & Tang, Q. (2018). Discrete cuckoo search algorithms for two-sided robotic assembly line balancing problem. *Neural Computing and Applications, 30*(9), 2685–2696.
15. Pham, D., & Karaboga, D. (2012). *Intelligent Optimisation Techniques: Genetic Algorithms, Tabu Search, Simulated Annealing and Neural Networks.* Springer Science & Business Media, London, UK.

5

Spatial Interrelation Shape Features

The spatial interrelation function defines the region or shape contour by the relation of its curves or pixels. The representation is completed utilizing its geometrical characteristics: length, curvature, relative location and orientation, distance, area, etc.

5.1 Adaptive Grid Resolution (AGR)

In the AGR, a grid that is square and large enough to fit the whole shape is covered with a shape [1]. The grid cell resolution diverges from part to part, depending on the content of the part of the form. The high resolution is applied to the contour or detail part of the shape, while low resolution is applied to the rough areas of the shape.

To ensure the invariance of orientation, one must translate a random oriented shape into an inimitable mutual orientation. Initially, search the main shape axis. The main axis is the straight-line segment, which links the two points on the outline the most distant from each other. The shape is then rotated so that the main axis parallels the x-axis. This rotation is not yet exclusive, because there are two choices: one point may be on the right or the left. This issue is resolved by calculating the polygon centroid and ensuring that the center of gravity is beneath the main axis, consequently ensuring an exclusive rotation.

The calculation technique for the shape's AGR representation uses quad-tree breakdown to the bitmap presentation of the shape [2]. When the bitmap is successfully subdivided into four equal size quadrants, the decomposition is created. If a bitmap quadrant doesn't contain part of the shape completely, it is recursively divided into smaller quadrants until the bitmap quadrants are reached, i.e. the recursion's end condition is attained at the predefined resolution value. Figure 5.1a is an AGR example.

To denote the AGR image, the quad-tree technique is imposed. Every quad-tree node covers a bitmap square area. The node level in the quad-tree detects the shape size. The interior nodes (exhibited by grey circles) characterize partly obscured areas; the white colored boxes signify the leaf node areas with all zeros whereas the black colored boxes denote the leaf nodes areas with all ones. The all ones areas are utilized to denote the shape that is illustrated in Figure 5.1b.

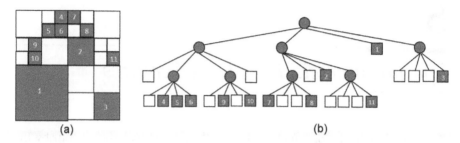

FIGURE 5.1
Adaptive resolution representation. (a) AGR image; (b) Quad-tree breakdown of AGR.

This method has several advantages and limitations listed as follows:

Advantages: Because the normalization is done prior to the computation of AGR, AGR depiction is not variant under scaling, orientation, and translation. It is also simple enough to compute.

Limitations: Setting the proper resolution for a particular shape is difficult.

5.2 Bounding Box (BB)

BB calculates homeomorphisms among its shape and two-dimensional lattices or frames. This projection is not limited to simply linked shapes, but unlike many other approaches [3], it applies to random topologies.

To create an invariant bounding box depiction to orientation, a shape should be normalized prior to further calculation. The smallest bounding box or bounding rectangle of shape S is represented by $B(S)$; its height and width are referred to as h and w, respectively.

The BB splits a shape S into p (row) $\times$ q (column) segments. An illustration of this process and its outcome is shown in Figure 5.2.

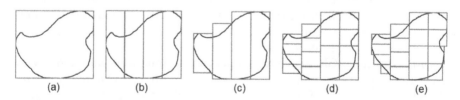

FIGURE 5.2
Five bounding box split steps. (a) Bounding box of a pixel set S; (b) Split S into q vertical parts; (c) Calculate the bounding box of every resultant set S_j where $j = 1, 2, ..., q$; (d) Split every bounding box into p horizontal parts; (e) Calculate the bounding box of every resultant pixel set S_{ij} where $i = 1, 2, ..., p$.

If $L = (L_x, L_y)^T$ signifies the bottom left corner position of the first bounding box of the shape S and $C_{ij} = (c_x^{ij}, c_y^{ij})$ represents the center of sample box, then the coordinates can be represented by equation (5.1).

$$\begin{pmatrix} \mu_x^{ij} \\ \mu_y^{ij} \end{pmatrix} = \begin{pmatrix} (c_x^{ij} - L_x)/w \\ (c_y^{ij} - L_y)/h \end{pmatrix} \tag{5.1}$$

This delivers a representation of S that is scale invariant. Sampling M points of a $p \times q$ lattice consequently allows characterization of S as a vector (v) as represented in equation (5.2).

$$v = \left[\mu_x^{i(1)j(1)}, \mu_y^{i(1)j(1)}, ..., \mu_x^{i(M)j(M)}, \mu_y^{i(M)j(M)} \right] \tag{5.2}$$

There are several advantages and limitations of BB that are listed here.

Advantages: Bounding box depiction is an easy computational geometry method to calculate homeomorphisms among lattices and shapes. It is efficient in time and storage. It is invariant for scaling, orientation, and translation and resistant to noisy shape contours.

Limitations: Setting the proper value of horizontal and vertical slices is challenging.

5.3 Convex Hull (CH)

A number of convex hulls are used to represent shape in this approach. The convex hull (CH) is the tiniest convex polygon containing an object entirely. The object boundary is smoothed before this approach is applied [4]. For a region S, the CH is demarcated as the minimum convex set comprising S.

A concavity tree (Figure 5.3) can be obtained as the shape depiction after applying this method recursively. Every concavity can be defined by the following:

- Area
- Chord length
- Extreme curvature
- Distance to the chord from extreme curvature point

The similarity checking among shapes is converted to a graph or a string matching.

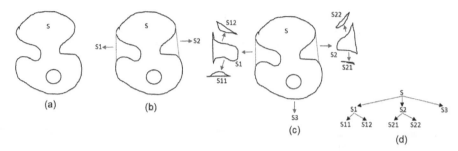

FIGURE 5.3
Recursive procedure of CH. (a) Original shape; (b) CH and its concavities: Step 1; (c) CH and its concavities: Step 2; (d) Concavity tree representation of CH.

This method has several advantages and limitations that are discussed as follows:

Advantages: CH depiction has a large storing efficacy. It is not variant for scaling, orientation, and translation and is also resistant to noisy shape contour (after filtering).

Limitations: Convex hulls are prone to errors when the images are not exactly same.

5.4 Chain Code (CC)

CC is a method to represent an object contour by a connected sequence of straight-line segments of specified direction and length. This straight-line segment is a sequence of integers [5].

5.4.1 Basic

Basic CC designates the motion across a digital curve or an arrangement of contour pixels by utilizing 4-connectivity or 8-connectivity [6]. Every movement direction is determined by numbers from 0 to 3 (in the case of 4-connectivity) or from 0 to 7 (in the case of 8-connectivity) signifying an angle of 90° or 45° separation in a counter clockwise direction concerning the positive *x*-axis, as represented in Figure 5.4.

This method is invariant to translation. The boundaries can be matched by equating their CCs, but there are two main limitations to this method: (1) it is variant to orientation; (2) it is very delicate to noise. To correct these issues, differential CC (DCC), re-sampling CC (RCC), and vertex CC (VCC) were proposed.

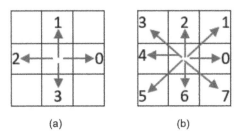

(a)   (b)

FIGURE 5.4
Direction of basic CC. (a) CC in four direction (4-connectivity); (b) CC in eight direction (8-connectivity).

5.4.2 Differential

DCC is attained from first difference of CC [7]. The first difference of CC is attained by considering two numbers of CC and computing the number of transitions needed to reach the second number in a counter clockwise direction from the first number. The first difference is orientation invariant. The shape number is attained after normalization of DCC. For the least magnitude of shape number, normalization is utilized. Normalization treats a CC as a circular sequence and redefines the initial point so that the resultant sequence of numbers comprises a least integer.

5.4.3 Re-sampling

RCC consists of re-sampling the edge onto a rougher grid and then calculating the CCs of this rougher presentation [8]. This eliminates minor discrepancies and noise but can help to recompense for differences in length of the chain code because of the pixel grid.

5.4.4 Vertex

VCC enhances the chain code profitability [9]. The VCC is based on the cell vertices numbers that are in touch with the shape's bounding edge. In the vertex chain code, only three elements are used to represent the shape 1 (if two vertices touch), 2 (if three vertices touch), and 3 (if four vertices touch). Figure 5.5 shows VCC to recognize the elements for shape representation.

5.4.5 Chain Code Histogram (CCH)

The CCH replicates the probabilities of various directions within the contour [10]. If CC is utilized to match, it must not be dependent on the initial pixel selection in the arrangement. CC is generally large in dimensions and delicate to distortion and noise. Therefore, excluding the CCH, the other CC methods are frequently utilized to represent the contour, but not as contour attributes. Figure 5.6 shows the CCH representation of an image.

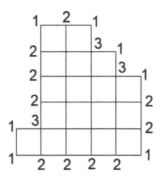

FIGURE 5.5
Representation of VCC.

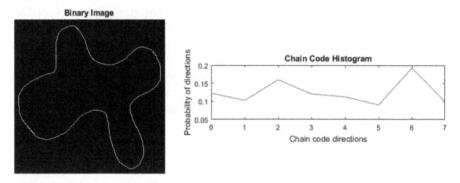

FIGURE 5.6
CCH representation of a binary image.

There are several advantages and limitations of chain code representation.

Advantages: A compressed contour representation.

Limitations: Chain code depends on the starting point. Operations such as scaling and rotation result in different contours that in practice cannot be normalized (due to a finite grid) and hence in different chain codes.

5.5 Smooth Curve Decomposition (SCD)

SCD can be considered as a shape descriptor [11]. The segments from a Gaussian smoothed boundary to the curvature zero-crossing points are utilized to attain primitives, called tokens. Every token's extreme curvature and rotation is considered a feature. Weighted Euclidean distance measurement is used to check the similarity among two tokens. The non-metric distance is used to measure the shape similarity.

There are several advantages and limitations of this method.

Advantages: Shape retrieval built on token representation is vigorous in the occurrence of partly occulted objects, orientation, scaling, and translation.

Limitations: For a specific shape, it is difficult to set the proper amount of Gaussian smoothing.

5.6 Beam Angle Statistics (BAS)

The shape descriptor for BAS is created on the beams initiated from a shape contour point [12,13]. Let S be the shape contour. $S = \{P_1, P_2, ..., P_M\}$ is denoted by a connected sequence of points, $P_i = (x_i, y_i)$, $i = 1, 2, ..., M$, where M is the number of contour points and $P_i = P_i + M$. For each point P_i, the beams are defined as the set of vectors as shown in equation (5.3).

$$L(P_i) = \{FV_{i+n}, BV_{i-n}\} \tag{5.3}$$

where $FV_{i+n} = \overrightarrow{P_i P_{i+n}}$ (forward vector) and $BV_{i-n} = \overrightarrow{P_i P_{i-n}}$ (backward vector) in the n-th order neighborhood system (see Figure 5.7, $n = 4$ for example). For every point P_i, the beam angle between FV_{i+n} and BV_{i-n} in the n-th order neighborhood system is represented by equation (5.4).

$$A_n(i) = (\theta_{FV_{i+n}} - \theta_{BV_{i-n}})$$

where :

$$\theta_{FV_{i+n}} = \arctan \frac{y_{i+n} - y_i}{x_{i+n} - x_i} \tag{5.4}$$

$$\theta_{BV_{i-n}} = \arctan \frac{y_{i-n} - y_i}{x_{i-n} - x_i}$$

For every contour point P_i, the beam angle $A_n(i)$ can be used as an arbitrary variable with the probability density function $P(A_n(i))$. Hence, for a shape descriptor, BAS may deliver a compact depiction. For this motive, the k-th moment of the arbitrary variable $A_n(i)$ is demarcated as shown in equation (5.5).

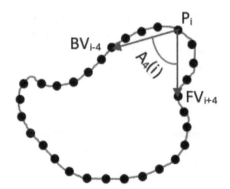

FIGURE 5.7
Beam angle at the 4-th order neighborhood system for a contour point.

$$E\left[A^k(i)\right] = \sum_{n=1}^{(M/2)-1} A_n^k(i).P_n\left(A_n(i)\right); k = 1, 2, \dots \tag{5.5}$$

Here E is the expected value.

There are several advantages and limitations of this method that are listed here.

Advantages: The shape descriptor for the BAS captures the perceptual information utilizing the statistical information on the discrete point's beam. It produces globally distinct characteristics at every point of contour by using the entire border points. The BAS descriptor is also relatively steady under inconsistencies and not variant for orientation, scaling, and translation.

Limitations: BAS can't calculate the feature vectors of boundary points in the lack of parametric border representation. Thus, it can't be utilized to calculate the feature vectors of edge pixels even if it is not possible to extract the shape boundary.

5.7 Shape Matrix (SM)

SM descriptor is a $P \times Q$ matrix to present a region shape. There are two basic modes of shape matrix: Square model and polar model.

5.7.1 Square Model

This method is also referred to as grid descriptor. Basically, a cell grid is covered in a shape and the grid is scanned from left to right and from top to bottom. The outcome is a bitmap. The cells covered with the shape are assigned to 1 and the ones not covered with the shape to 0. The shape can then be shown as a binary function vector [14].

Split the square into $Q \times Q$ sub squares and represent by S_{ni}, $n, i = 1,..., Q$, the sub squares of the created grid. Express the shape matrix B_{ni} as shown in equation (5.6).

$$B_{ni} = \begin{cases} 1 \Leftrightarrow & \mu(S_{ni} \cap S) \geq \mu(S_{ni})/2 \\ 0 & \text{otherwise} \end{cases} \tag{5.6}$$

where S is the shape and μ is the area of the shape. Figure 5.8 demonstrates the method.

For a shape with more than one maximum radius, several shape matrices can be defined, and the distance of similarity between these matrices is the smallest distance.

5.7.2 Polar Model

The polar model of a shape matrix is characterized by the following steps [15]. A polar raster of concentric circles and radial lines is overlaid in the centroid of the shape (Figure 5.9a). The binary shape value is measured at the intersections of the circles (p) and radial lines (q). The shape matrix is formed in such a way that the circles match the matrix columns and the radial lines match the matrix rows. Figure 5.9 demonstrates an instance of the polar shape model (PSM), where $p = 4$ and $q = 8$.

This representation is easier than the square model as it only utilizes one matrix regardless of the number of the largest radii of the shape. However, as the sampling density with the polar sampling raster is not static, a weighed shape matrix is needed.

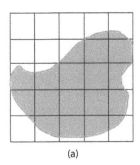

$$B = \begin{bmatrix} 0 & 0 & 0 & 0 & 0 \\ 0 & 0 & 1 & 1 & 1 \\ 1 & 1 & 1 & 1 & 1 \\ 1 & 1 & 1 & 1 & 1 \\ 0 & 1 & 1 & 1 & 1 \end{bmatrix}$$

(a)   (b)   (c)

FIGURE 5.8
Square model shape matrix: (a) Original shape region; (b) Square model shape matrix; (c) Rebuilding of shape region.

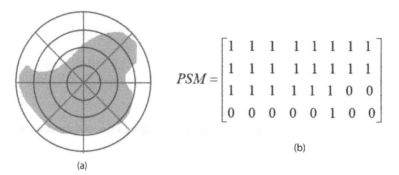

$$PSM = \begin{bmatrix} 1 & 1 & 1 & 1 & 1 & 1 & 1 & 1 \\ 1 & 1 & 1 & 1 & 1 & 1 & 1 & 1 \\ 1 & 1 & 1 & 1 & 1 & 1 & 0 & 0 \\ 0 & 0 & 0 & 0 & 0 & 1 & 0 & 0 \end{bmatrix}$$

(b)

(a)

FIGURE 5.9
Polar shape model: (a) Shape region; (b) Polar shape model of the shape.

There are several advantages and limitations of this method which are presented here.

Advantages: The shape matrix for each compact shape is available. The size of the shapes that can be characterized by the shape matrix is not destined. Further, shape matrix can be designed for a shape with holes. The polar matrix representation is invariant to translation, rotation, and scaling. The object can be recreated from the shape matrix.

Limitations: The accuracy of shape recognition by using this method is specified by the grid cell's size. Thus, if the size of the grid is not proper with respect to the shape, then the recognition may not be accurate.

5.8 Shape Context (SC)

SC has been revealed to be a useful tool for the tasks of object recognition [16]. A shape is represented by a discrete set of sampled points from the shape's internal or external contours. These can be obtained as edge pixel locations $P = \{p_1, p_2, ..., p_n\}$ of n points as found in an edge detector. The set of vectors is calculated from one point (p_i) to all the other shape sample points. These vectors express the configuration of the whole shape in relation to the point of reference. This set of $n-1$ vectors is obviously a rich description, because as n grows big, the representation of the shape becomes accurate. For a shape contour point p_i, a course histogram of the relative coordinated of the remaining $n-1$ points is computed. This histogram is referred to as the shape context of p_i.

There are several advantages and limitations of SC which are presented here.

Advantages: SC has been useful to a variability of object recognition issues. The shape context descriptor has the different properties of invariance. This descriptor is fundamentally invariant to translation because it is built on relative point positions. For clutter-free images, the scale invariant descriptor can be obtained by standardizing the radial distances by the median (or mean) distance between entire points of pairs. The rotation invariant descriptor can be formed by orienting the coordinate system at every point in order to align the positive x-axis with the tangent vector. The descriptor is vigorous against small discrepancies of the shape. Outliers are the points with an ultimate matching cost greater than a threshold value. To reduce the outlier's effects, dummy points are added.

Limitations: The performance of the shape context algorithm is strictly affected by the number of sample points selected.

5.9 Chord Distribution (CD)

The main goal of this descriptor is to compute all the shape's chord lengths (entire pair-wise distances between contour points) and to create a histogram of their lengths and rotations. Figure 5.10 shows an example of CD [17].

There are some advantages and limitations of CD which are presented here.

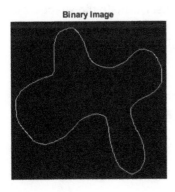
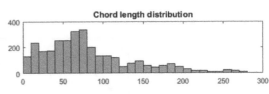

FIGURE 5.10
Chord distribution.

Advantages: The length histogram is not variant to scales and rotation linearly with the object's size. The chord angle histogram is not variant to object size and shifts comparative to orientation of the object. Also, because it is fast, it is suitable for solving an initial alignment transformation between shapes, and it can be used for matching, indexing, retrieval, and recognition of shapes in large databases.

Limitations: These descriptors are not satisfactorily compact. Furthermore, they rely constantly on a reference point whereby the shape border is parameterized. This dependence is easy as the contour is closed, and any point on the contour can be utilized as a reference point, thus these features might be changed.

5.10 Shock Graphs (SG)

SGs are based on a medial axis [18]. A skeleton can be defined as a linked set of medial lines on a figure's limbs. For instance, the skeleton can be the path actually traveled by the pen in the case of thick hand-drawn characters. In reality, the basic concept of the skeleton is to eliminate redundant information while only retaining topological information about the structure of the object, which can help to recognize it. The medial axis is the locus of the maximum disks centers that fit in the shape. This is shown in Figure 5.11. The bold line in the figure is a rectangular shaded skeleton. The skeleton can then be broken down into segments and graphically represented according to certain criteria.

SG is a concept of the shape that breaks a shape into a group of hierarchically prepared primitive portions. Shock segments are monotonic flow curved sequences of the medial axis and deliver the medial axis segments in a more sophisticated way. This descriptor can discriminate the shapes, but the medial axis can't. The matching between shapes becomes a graph matching. Figure 5.12 displays a shape and its shock graph.

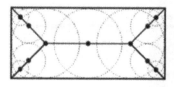

FIGURE 5.11
Medial axis of a rectangle.

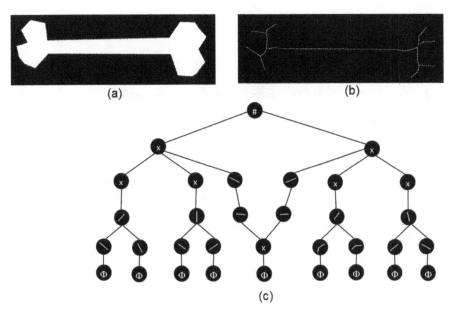

FIGURE 5.12
(a) Binary image; (b) Medial axis segment of (a); (c) Shock graph representation of (b).

There are some advantages and limitations of SG that are presented here.

Advantages: The algorithm of shock graphs has decent performance for contour disturbances, parts articulation and deformation, scale variations, segmentation inconsistencies, viewpoint varieties, and partial occultation.

Limitations: The computation complexity is very high.

5.11 Summary

Spatial feature descriptor is a shape description technique. These descriptors can use statistical theory (beam angle statistics, chain code histogram, chord distribution and shape context), tree-based analysis (convex hull and adaptive grid resolution), or syntactic analysis (smooth curve decomposition) to extract or characterize a shape characteristic. This structure of description not only compresses the shape data, but also delivers an expressive and compact form to promote additional operations of recognition.

References

1. Karaa, W. B. A., & Dey, N. (2017). *Mining Multimedia Documents*. CRC Press, Boca Raton, FL.
2. Yang, X., & Tian, Y. (2014). Super normal vector for activity recognition using depth sequences. In *Proceedings of the IEEE Conference on Computer Vision and Pattern Recognition* (pp. 804–811).
3. Chaki, J., & Dey, N. (2018). *A Beginner's Guide to Image Preprocessing Techniques*. CRC Press, Boca Raton, FL.
4. Kotyk, T., Dey, N., Ashour, A. S., Balas-Timar, D., Chakraborty, S., Ashour, A. S., & Tavares, J. M. R. (2016). Measurement of glomerulus diameter and Bowman's space width of renal albino rats. *Computer Methods and Programs in Biomedicine, 126,* 143–153.
5. Dey, N., Ashour, A. S., & Hassanien, A. E. (2017). Feature detectors and descriptors generations with numerous images and video applications: A recap. In *Feature Detectors and Motion Detection in Video Processing*, Nilanjan Dey, Amira Ashour and Prasenjit Kr. Patra (eds.), (pp. 36–65). IGI Global, Hershey, Pennsylvania.
6. Das, N., Pramanik, S., Basu, S., Saha, P. K., Sarkar, R., Kundu, M., & Nasipuri, M. (2014). Recognition of handwritten Bangla basic characters and digits using convex hull based feature set. arXiv preprint arXiv:1410.0478.
7. Shen, S., Bui, A. A., Cong, J., & Hsu, W. (2015). An automated lung segmentation approach using bidirectional chain codes to improve nodule detection accuracy. *Computers in Biology and Medicine, 57,* 139–149.
8. Cormier, O. S. M., & Ferrie, F. P. (2016, June). Evaluation of shape description metrics applied to human silhouette tracking. In *2016 13th Conference on Computer and Robot Vision (CRV)* (pp. 370–375). IEEE.
9. Raj, Y. A., & Alli, P. (2018). Pattern-based Chain Code for Bi-level Shape Image Compression, *Taga Journal of Graphic Technology, 14,* 3069–3080.
10. Gaber, T., Hassanien, A. E., El-Bendary, N., & Dey, N. (eds.). (2015). *The 1st International Conference on Advanced Intelligent System and Informatics (AISI2015)*, November 28–30, (Vol. 407). Springer, Beni Suef, Egypt.
11. Araki, T., Ikeda, N., Dey, N., Acharjee, S., Molinari, F., Saba, L., Godia, E. C., Nicolaides, A., & Suri, J. S. (2015). Shape-based approach for coronary calcium lesion volume measurement on intravascular ultrasound imaging and its association with carotid intima-media thickness. *Journal of Ultrasound in Medicine, 34*(3), 469–482.
12. Mahdikhanlou, K., & Ebrahimnezhad, H. (2014, May). Plant leaf classification using centroid distance and axis of least inertia method. In *2014 22nd Iranian Conference on Electrical Engineering (ICEE)* (pp. 1690–1694). IEEE.
13. Dey, N., Ashour, A. S., Shi, F., & Balas, V. E. (2018). *Soft Computing Based Medical Image Analysis*. Academic Press, London, UK.
14. Dey, N., Nandi, B., Das, P., Das, A., & Chaudhuri, S. S. (2013). Retention of electrocardiogram features insignificantly devalorized as an effect of watermarking for. *Advances in Biometrics for Secure Human Authentication and Recognition,* 175–212.

15. Araki, T., Ikeda, N., Dey, N., Acharjee, S., Molinari, F., Saba, L., Godia, E. C., Nicolaides, A., & Suri, J. S. (2015). Shape-based approach for coronary calcium lesion volume measurement on intravascular ultrasound imaging and its association with carotid intima-media thickness. *Journal of Ultrasound in Medicine, 34*(3), 469–482.

16. Chaki, J., Parekh, R., & Bhattacharya, S. (2015). Plant leaf recognition using texture and shape features with neural classifiers. *Pattern Recognition Letters, 58,* 61–68.

17. Elmezain, M., & Abdel-Rahman, E. O. (2015). Human activity recognition: Discriminative models using statistical chord-length and optical flow motion features. *Applied Mathematics & Information Sciences, 9*(6), 3063.

18. Chaki, J. (2018, January). An efficient two-stage Palmprint recognition using Frangi-filter and 2-component partition method. In *2018 Fifth International Conference on Emerging Applications of Information Technology (EAIT)* (pp. 1–5). IEEE.

6

Moment Shape Feature

Moment invariants are often used as image processing, shape recognition, remote sensing, and classification features. Moments can describe characteristics of uniquely shaped objects [1].

6.1 Contour Moment (CM)

CM, the study of a boundary or outline, can be utilized to decrease the dimension of contour depiction [2]. Assuming a shape contour with B contour points is denoted as a 1-dimensional shape representation $s(i)$, the p-th moment M_p and central moment μ_p can be assessed by equation (6.1).

$$M_p = \frac{1}{B}\sum_{i=1}^{B}\left[s(i)\right]^p$$

$$\mu_p = \frac{1}{B}\sum_{i=1}^{B}\left[s(i)-M_1\right]^p \tag{6.1}$$

The normalized moments are not variant to shape orientation, scaling and translation can be expressed by equation (6.2).

$$M_p{'} = \frac{M_p}{\mu_2^{\frac{p}{2}}}$$

$$\mu_p{'} = \frac{\mu_p}{\mu_2^{\frac{p}{2}}} \tag{6.2}$$

Shape descriptors with minimum noise-sensitivity can be attained from equation (6.3).

$$F_1 = \frac{(\mu_2)^{\frac{1}{2}}}{M_1}, F_2 = \frac{(\mu_3)}{(\mu_2)^{\frac{3}{2}}}, F_3 = \frac{(\mu_4)}{(\mu_2)^2} \tag{6.3}$$

The additional contour moments technique treats the 1D shape feature function $s(i)$ as an arbitrary variable a and generates an N bins histogram $H(a_i)$ from $s(i)$. Then, the p-th central moment is attained by equation (6.4).

$$\mu_p = \sum_{i=1}^{N} (a_i - M)^p H(a_i)$$

(6.4)

$$M = \sum_{i=1}^{N} a_i H(a_i)$$

There are several advantages and limitations of the method which are mentioned as follows:

Advantages: It is easy to develop.

Limitations: It is hard to develop moments of higher order with physical understanding.

6.2 Geometric Invariant Moment (GIM)

The GIM descriptor is the easiest of the moment functions with basis $\psi_{mn} = x^m y^n$; although it is complete, it is not orthogonal [3]. Geometric moment function G_{mn} of order $(m + n)$ is specified as shown in equation (6.5).

$$G_{mn} = \sum_{x} \sum_{y} x^m y^n I(x,y)$$

(6.5)

where $I(x, y)$ is the digital image function and $m, n = 0, 1, 2, \ldots$

The geometric central moments, which are not variant to translation, are represented as shown in equation (6.6) where x' and y' are the centroid

$$\mu_{mn} = \sum_{x} \sum_{y} (x - x')^m (y - y')^n I(x,y)$$

(6.6)

where $x' = \dfrac{G_{10}}{G}, y' = \dfrac{G_{01}}{G}$

The moments are additionally normalized for the effects of alteration of scale utilizing equation (6.7) [4].

$$\eta_{mn} = \frac{\mu_{mn}}{\mu_{00}^{\gamma}}$$

(6.7)

where
$$\gamma = \frac{m+n}{2} + 1$$

A group of seven geometric invariant moments (φ) derived from second and third order moments can be represented by equation (6.8).

$$\varphi_1 = \eta_{20} + \eta_{02}$$

$$\varphi_2 = (\eta_{20} - \eta_{02})^2 + 4\eta_{11}^2$$

$$\varphi_3 = (\eta_{30} - 3\eta_{12})^2 + (3\eta_{21} - \eta_{03})^2$$

$$\varphi_4 = (\eta_{30} + \eta_{12})^2 + 3(\eta_{21} + \eta_{03})^2$$

$$\varphi_5 = (\eta_{30} - 3\eta_{12})(\eta_{30} + \eta_{12})\left[(\eta_{30} + \eta_{12})^2 - 3(\eta_{21} + \eta_{03})^2\right] \qquad (6.8)$$

$$+ (3\eta_{21} - \eta_{03})(\eta_{21} + \eta_{03})\left[3(\eta_{30} + \eta_{12})^2 - (\eta_{21} + \eta_{03})^2\right]$$

$$\varphi_6 = (\eta_{20} - 3\eta_{02})\left[(\eta_{30} + \eta_{12})^2 - (\eta_{21} + \eta_{03})^2\right] + 4\eta_{11}(\eta_{30} + \eta_{12})(\eta_{21} + \eta_{03})$$

$$\varphi_7 = (3\eta_{21} - \eta_{03})(\eta_{30} + \eta_{12})\left[(\eta_{30} + \eta_{12})^2 - 3(\eta_{21} + \eta_{03})^2\right]$$

$$+ (3\eta_{21} - \eta_{03})(\eta_{21} + \eta_{03})\left[3(\eta_{30} + \eta_{12})^2 - (\eta_{21} + \eta_{03})^2\right]$$

Geometric invariant moments have several advantages and drawbacks that are listed as follows [5]:

Advantages: They are algorithmically simple. Furthermore, they are not variant to scaling, rotation, translation, and mirroring.

Limitations: The basis of this method is not orthogonal; thus, these moments suffer from high redundancy of information. Moments in higher order are very delicate to noise. The basis contains powers of m and n, thus, the moments calculated have huge discrepancy in the dynamic values range for various orders. This may cause numerical uncertainty when the image is big.

6.3 Zernike Moment (ZM)

ZMs are orthogonal moments [6]. The complex ZMs are obtained from orthogonal Zernike polynomials. The image is mapped into polar co-ordinates (r, θ) and onto unit circle. r is the length of the vector from the

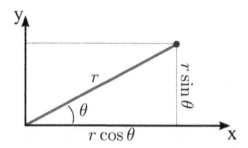

FIGURE 6.1
Polar coordinate representation of (x, y).

origin to the pixel (x, y); θ is the angle between the vector r and x axis in counter clockwise direction. (x, y) can be represented by polar coordinate (Figure 6.1) as shown in equation (6.9).

$$x = r\cos\theta, y = r\sin\theta \qquad (6.9)$$

The orthogonal radial polynomial $R_{pq}(r)$ is defined by equation (6.10).

$$R_{pq}(r) = \sum_{i=0}^{(p-|q|)/2} (-1)^i \frac{(p-i)!}{i! \times \left(\dfrac{p-2i+|q|}{2}\right)! \left(\dfrac{p-2i-|q|}{2}\right)!} r^{p-2i} \qquad (6.10)$$

$p = 0, 1, 2, \ldots; 0 \le |q| \le p$; and $p-|q|$ is even.

Zernike polynomials are a whole group of complex valued functions perpendicular over the unit disk, i.e., $x^2 + y^2 \le 1$. ZM of order p with duplication q of shape region $I(x, y)$ is specified by equation (6.11) [7].

$$Z_{pq} = \frac{p+1}{\pi} \sum_{r} \sum_{\theta} I(r\cos\theta, r\sin\theta) \cdot R_{pq}(r) \cdot \exp(jq\theta) \qquad (6.11)$$

where $r \le 1$, $j = \sqrt{-1}$ and $\theta = \tan^{-1}y/x$.

If the image is oriented by an angle α, the transformed ZM functions Z'_{pq} are represented by equation (6.12).

$$Z'_{pq} = Z_{pq} \cdot e^{-jq\alpha} \qquad (6.12)$$

This represents that the magnitude of the moments remains identical after the rotation [8].

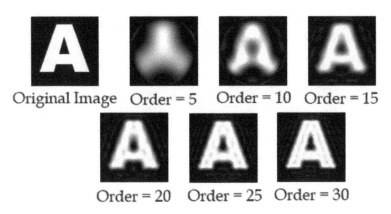

FIGURE 6.2
Reconstruction of an image with ZM.

Reconstruction of an image with ZM is shown in Figure 6.2.
ZMs have several advantages and limitations that are mentioned here.

Advantages: The magnitudes of ZMs are orientation invariant. They are vigorous to minor variations and noise in shape. They have the smallest redundancy of information because the basis is orthogonal.

Limitations: The image coordinate space essentially can be changed to the orthogonal polynomial domain demarcated. Discrete summations are computed from the continuous integrals. Not only does this calculation generate numerical inaccuracies in the calculated moments, but it also strictly affects quantitative features such as orthogonality and invariance of orientation. As the radial Zernike polynomial order grows into a large number, the computational complexity also rises.

6.4 Radial Chebyshev Moment (RCM)

RCM of order m and repetition n is represented by using equation (6.13). (r, θ) is the polar coordinate representation where r is the length of the vector from the origin to the pixel (x, y); θ is the angle between the vector r and x axis in a counter clockwise direction [9].

$$C_{mn} = \frac{1}{2\pi\rho(m,p)} \sum_{r=0}^{p-1} \sum_{\theta=0}^{2\pi} t_m(r) \cdot \exp(-jn\theta) \cdot I(r,\theta) \tag{6.13}$$

For an image I of size $D \times D$, $t_m(r)$ is the scaled orthogonal Chebyshev polynomials which can be denoted by equation (6.14).

$$t_0(x) = 1$$

$$t_1(x) = \frac{(2x - D + 1)}{D} \qquad (6.14)$$

$$t_m(x) = \frac{(2m-1)t_1(x)t_{m-1}(x) - (m-1)\left\{1 - \frac{(m-1)^2}{D^2}\right\}t_{m-2}(x)}{m}, m > 1$$

The squared-norm $\rho(m, D)$ is represented by equation (6.15).

$$\rho(m, D) = \frac{D\left(1 - \frac{1}{D^2}\right)\left(1 - \frac{2^2}{D^2}\right)\cdots\left(1 - \frac{m^2}{D^2}\right)}{2m + 1} \qquad (6.15)$$

where $m = 0, 1, \ldots, D-1$ and $p = (D/2) + 1$.

The mapping among image coordinates (x, y) and (r, θ) is represented by equation (6.16).

$$x = \frac{rD}{2(p-1)}\cos(\theta) + \frac{D}{2}$$

$$y = \frac{rD}{2(p-1)}\sin(\theta) + \frac{D}{2} \qquad (6.16)$$

Radial Chebyshev moments have several advantages and limitations that are mentioned here.

Advantages: This method has rotational invariance property.

Limitations: This moment samples the identical number of points for various radii when computing the values of the moments, which leads to too many sampling points when the radius is small and inadequate sampling points when the radius is large. So, this moment can't attain decent performance in calculating efficacy and calculating accuracy concurrently; particularly for large images, the moments don't work well.

6.5 Legendre Moment (LM)

Moments with Legendre polynomials as kernel function are called Legendre moments [10]. LMs fit to the class of orthogonal moments, and they are utilized in numerous pattern recognition applications.

The two-dimensional LM L_{mn} of order $(m + n)$, with digital image function $I(x, y)$, demarcated on the square $[-1, 1] \times [-1, 1]$ is represented by equation (6.17).

$$L_{mn} = \frac{(2m+1)(2n+1)}{4} \int_{-1}^{1}\int_{-1}^{1} P_m(x)P_n(y)I(x,y)dxdy \qquad (6.17)$$

where the Legendre polynomial, $P_m(x)$, of order m is denoted by equation (6.18).

$$P_m(x) = \sum_{i=0}^{m} \left\{ (-1)^{\frac{m-i}{2}} \frac{1}{2^m} \frac{(m+i)!x^i}{\left(\frac{m-i}{2}\right)!\left(\frac{m+i}{2}\right)!i!} \right\}_{m-i=even} \qquad (6.18)$$

The recurrence relationship of Legendre polynomials, $P_m(x)$ of order m, is given as shown in equation (6.19).

$$P_m(x) = \frac{(2m-1)xP_{m-1}(x) - (m-1)P_{m-2}(x)}{m} \qquad (6.19)$$

where $P_0(x) = 1$, $P_1(x) = x$ and $p > 1$. As the area of Legendre polynomial [9] is the interior of $[-1, 1]$, a square image of $D \times D$ pixels with intensity function $I(s, t)$, $0 \le s, t \le (D-1)$, is resized in the region of $-1 < x, y < 1$. The outcome of equation (17) can now be depicted in discrete form as shown in equation (6.20).

$$L_{mn} = \lambda_{mn} \sum_{s=0}^{D-1}\sum_{t=0}^{D-1} P_m(x_s)P_n(y_t)I(s,t) \qquad (6.20)$$

where the normalizing constant is denoted by equation (6.21),

$$\lambda_{mn} = \frac{(2m+1)(2n+1)}{D^2} \qquad (6.21)$$

x_s and y_t represent the normalized pixel coordinates in the range of $[-1, 1]$, which are represented by equation (6.22).

$$x_s = \frac{2s}{D-1} - 1, y_t = \frac{2t}{D-1} - 1 \qquad (6.22)$$

This method has several advantages and limitations that are listed as follows:

> *Advantages:* This moment can be used in image reconstruction. This moment can be utilized to reach a near zero value of redundancy measure in a group of moment functions, so that the moments resemble independent features of the image.
>
> *Limitations:* This method cannot convert into a rotational invariant form.

6.6 Homocentric Polar-Radius Moment (HPRM)

The contour of the object is attained using an 8 connected neighbor area edge algorithm [11]. It is a single-pixel wide contour. The function of polar-radius is as defined in equation (6.23).

$$r_p = \sqrt{\left(x_p - x_c\right)^2 + \left(y_p - y_c\right)^2} \tag{6.23}$$

where (x_p, y_p) are the coordinates of the p-th contour pixel. The centroid of the object contour is (x_c, y_c). $P = 0, 1, \ldots, M-1$, where M is the overall number of the contour pixels.

$$x_c = \frac{1}{M}\sum_{p=0}^{M-1}x_p, \; y_c = \frac{1}{M}\sum_{p=0}^{M-1}y_p \tag{6.24}$$

By this technique, the function of polar-radius can be attained.

The polar-radius moment is defined as shown in equation (6.25).

$$\mu_m = \sum_{p=0}^{M-1} r_p^m \tag{6.25}$$

where μ_m is the m-th order moment of the contour.

The normalized moment and central normalized moment are represented by equation (6.26).

$$\mu_{nm} = \frac{1}{M}\sum_{p=0}^{M-1}\left(\frac{r_p}{\bar{r}}\right)^m, \; \mu_{ncm} = \frac{1}{M}\sum_{p=0}^{M-1}\left(\frac{r_p - \bar{r}}{\bar{r}}\right)^m \tag{6.26}$$

where $\bar{r} = \dfrac{1}{M}\sum_{p=0}^{M-1}r_p$

Let r_{max} and r_{min} be the longest and shortest one in entire r_p. The radius R_n of the n-th circle is defined by equation (6.27).

$$R_n = r_{min} + \left(\frac{m}{t} \times W \right) \tag{6.27}$$

where $W = r_{max} - r_{min}$, $m = 1, 2, ..., t$, W is the width, and t is the total number of the circles. $\mu_{ncp}(R_n)$ is the shape descriptor using a homocentric polar-radius moment.

HPRM has several advantages and limitations that are mentioned here.

Advantages: The method has the invariance properties against translation, rotation, and scaling.

Limitations: Selecting the effective value of t is difficult.

6.7 Orthogonal Fourier-Mellin Moment (OFMM)

OFMMs are expressed in a polar coordinate system (r, θ) over the unit circle as shown in equation (6.28) [12].

$$FM_{pq} = \frac{1}{2\pi\alpha_p} \int_0^{2\pi} \int_0^1 I(r,\theta) H_p(r) \exp(-jq\theta) r\, dr\, d\theta \tag{6.28}$$

where the rounded harmonic order $q = 0, \pm1, \pm2, ...$, and the $H_p(r)$ is a radial polynomial in r of degree p. The polynomials $H_p(r)$ are expressed by equation (6.29) over the $0 \leq r \leq 1$.

$$H_p(r) = \sum_{i=0}^{p} \alpha_{pi} r^i \tag{6.29}$$

where $\alpha_{pi} = (-1)^{p+i} \dfrac{(p+i+1)!}{(p-i)!i!(i+1)!}$

Hence, the normalization constant in equation (6.28) is expressed by equation (6.30).

$$\alpha_p = \frac{1}{2(p+1)} \tag{6.30}$$

Here α_{pi} is a coefficient of the p-th polynomial with p initiating from zero. The radial polynomial $H_p(r)$ and the harmonic polynomial $\exp(-jq\theta)$ have unrelated variables r and θ, and parameters p and q where $j = \sqrt{-1}$.

The radial polynomial $H_p(r)$ can be computed recursively as well and expressed by equation (6.31).

$$H_0(r) = 1,$$

$$H_1(r) = -2 + 3r,$$

$$H_p(r) = \frac{\left(2r\left(4p^2 - 1\right) - 4p^2\right)H_{p-1}(r) - (p-1)(2p+1)H_{p-2}(r)}{(p+1)(2p-1)}$$

(6.31)

As the group of $H_p(r)$ is orthogonal over the range $0 \le r \le 1$:

$$\int_0^1 H_p(r)H_k(r)r\,dr = \alpha_p \delta_{pk}$$

(6.32)

where δ_{pk} is the Kronecker symbol; the basis function $H_p(r)\exp(-jq\theta)$ in equation (6.28) is orthogonal over the unit circle.

One of the vital features of the orthogonal Fourier-Mellin moment is its rotating invariance [13]. If the image function $I(r, \theta)$ is oriented by an angle α and its moments after orientation are denoted as FM'_{pq}, the relationship between FM'_{pq} and FM_{pq} is expressed by equation (6.33).

$$FM'_{pq} = FM_{pq}\exp(-jq\theta)$$

(6.33)

which designates that the magnitude of the orthogonal Fourier-Mellin moments, $|FM_{pq}|$, is an orientation-invariant feature of the image function $I(r, \theta)$.

The orthogonality of the group $H_p(r)\exp(-jq\theta)$ permits us to recreate an image function demarcated in the unit circle by the inverse orthogonal Fourier-Mellin transform as shown in equation (6.34).

$$I(r,\theta) = \sum_{p=0}^{\infty} \sum_{q=-\infty}^{\infty} FM_{pq}H_p(r)\exp(jq\theta)$$

(6.34)

In practice, with a limited group of orthogonal Fourier-Mellin moments FM_{pq}, where—$Q_{max} \le q \le Q_{max}$ and $0 \le p \le P_{max}$, a rough version of $I(r, \theta)$, $I'(r, \theta)$ can be attained by equation (6.35).

$$I'(r,\theta) = \sum_{p=0}^{P_{max}} \sum_{q=-Q_{max}}^{Q_{max}} FM_{pq}H_p(r)\exp(jq\theta)$$

(6.35)

The following are some advantages and limitations of OFMM.

Advantages: This moment is orientation invariant. For small images, the description by OFMM is better than that by the Zernike moments in terms of image reconstruction errors and signal-to-noise ratio.

Limitations: This method is very sensitive to noise. OFMMs suffer from geometric error and numerical integration error. The geometric error rises when the square image is mapped into a unit disk and the mapping doesn't become perfect. The numerical integration error arises when the double integration comes close to the zeroth order summation.

6.8 Pseudo-Zernike Moment (PZM)

The kernel of PZM is a group of orthogonal pseudo-Zernike polynomials demarcated over the polar coordinate space in a unit circle [14]. The two-dimensional pseudo-Zernike moments of order m with repetition n of an image intensity function $I(r, \theta)$ are defined as shown in equation (6.36).

$$Z_{mn} = \frac{m+1}{\pi} \int_{-\pi}^{\pi} \int_{0}^{1} P_{mn}^{*}(r,\theta) I(r,\theta) r dr d\theta; |r| \le 1 \tag{6.36}$$

where pseudo-Zernike polynomials $P_{mn}(r,\theta)$ are represented by equation (6.37),

$$P_{mn}(r,\theta) = R_{mn}(r) e^{jq\theta}; j = \sqrt{-1} \tag{6.37}$$

and the real-valued radial polynomials, $R_{mn}(r)$, are specified as in equation (6.38).

$$R_{mn}(r) = \sum_{i=0}^{m-|n|} (-1)^{i} \frac{(2m+1-i)!}{i!(m+|n|+1-i)!(m-|n|-i)!} r^{m-i} \tag{6.38}$$

where $0 \le |n| \le m$. As PZMs are demarcated [15] in terms of polar coordinates (r, θ) with $|r| \le 1$, the calculation of pseudo-Zernike polynomials needs a linear transformation of the image coordinates (p, q), $p, q = 0, 1, 2,..., T-1$ to a suitable domain $(x, y) \in R^2$ inside a unit circle. Based on this figure, the subsequent discrete approximation of the continuous pseudo-Zernike moments' integral in equation (6.36) can be represented by equation (6.39).

$$Z_{mn} = \lambda(m,T) \sum_{p=0}^{T-1} \sum_{q=0}^{T-1} R_{mn}(r_{mn}) e^{-jn\theta pq} I(p,q); 0 \le r_{pq} \le 1 \qquad (6.39)$$

where the image coordinate transformation to the domain of the unit circle is presented by equation (6.40).

$$r_{pq} = \sqrt{x_p^2 + y_q^2}; \theta_{pq} = \tan^{-1}\left(\frac{y_q}{x_p}\right); \lambda(m,T) = \frac{m+1}{(T-1)^2}$$

$$x_p = c_1 p + c_2; y_p = c_1 q + c_2; c_1 = \frac{2}{T-1}; c_2 = -1 \qquad (6.40)$$

The image intensity function $I(p, q)$ can be recreated from a limited number S orders of pseudo Zernike moments using equation (6.41).

$$I(p,q) \approx \frac{Z_{m0}}{2} R_{m0} + \sum_{m=0}^{S} \sum_{n} \left[Z_{mn}^c \cos(n\theta_{pq}) + Z_{mn}^s \sin(n\theta_{pq}) \right] R_{mn}(r_{pq}) \qquad (6.41)$$

where Z_{mn}^c and Z_{mn}^s represent the real and imaginary parts of Z_{mn}, respectively. The followings are some of the advantages and limitations of PZM.

Advantages: PZMs have been proven to be superior to other moment functions such as Zernike moments in terms of their feature representation abilities. PZMs offer more feature vectors than Zernike moments.

Limitations: Computation time is high.

6.9 Summary

Moment features are used for the recognition of a small group of distinct objects. This feature can be used for pre-classification for more accurate comparison (boundary matching) in order to decrease the number of candidates. Typical applications using moment features are image recognition, robot vision, character recognition, etc. Simple operations are involved to compute moment features. This type of feature can be computed from image matrix, run-length code, and chain code. Boundary-based moment shape descriptors consider only the boundary information, overlooking the shape's internal content. Thus, these moment descriptors can't signify shapes for which the whole boundary information is not accessible. On the other hand, region-based moment descriptors exploit both boundary and interior pixels, and hence are relevant to generic shapes.

References

1. Karakasis, E. G., Amanatiadis, A., Gasteratos, A., & Chatzichristofis, S. A. (2015). Image moment invariants as local features for content based image retrieval using the bag-of-visual-words model. *Pattern Recognition Letters, 55*, 22–27.

2. Chaki, J., Parekh, R., & Bhattacharya, S. (2015). Plant leaf recognition using texture and shape features with neural classifiers. *Pattern Recognition Letters, 58*, 61–68.

3. Ghosh, A., Sarkar, A., Ashour, A. S., Balas-Timar, D., Dey, N., & Balas, V. E. (2015). Grid color moment features in glaucoma classification. *International Journal of Advanced Computer Science and Applications, 6*(9), 1–14.

4. Chaki, J., Parekh, R., & Bhattacharya, S. (2018). Plant leaf classification using multiple descriptors: A hierarchical approach. *Journal of King Saud University-Computer and Information Sciences.*

5. Samanta, S., Ahmed, S. S., Salem, M. A. M. M., Nath, S. S., Dey, N., & Chowdhury, S. S. (2015). Haralick features based automated glaucoma classification using back propagation neural network. In *Proceedings of the 3rd International Conference on Frontiers of Intelligent Computing: Theory and Applications (FICTA) 2014* (pp. 351–358). Springer, Cham, Switzerland.

6. Le-Tien, T., Huynh-Kha, T., Pham-Cong-Hoan, L., Tran-Hong, A., Dey, N., & Luong, M. (2017). Combined zernike moment and multiscale analysis for tamper detection in digital images. *Informatica, 41*(1), 59–70.

7. Dey, N., Ashour, A. S., Shi, F., & Balas, V. E. (2018). *Soft Computing Based Medical Image Analysis.* Academic Press, Cambridge, UK.

8. Ammar, M., Mahmoudi, S., & Stylianos, D. (2018). A set of texture-based methods for breast cancer response prediction in neoadjuvant chemotherapy treatment. In *Soft Computing Based Medical Image Analysis* (pp. 137–147). Academic Press, Cambridge, Massachusetts.

9. Flusser, J., Suk, T., & Zitová, B. (2016). *2D and 3D Image Analysis by Moments.* John Wiley & Sons, Chichester, UK.

10. Gospodinova, E., Gospodinov, M., Dey, N., Domuschiev, I., Ashour, A. S., & Sifaki-Pistolla, D. (2015). Analysis of heart rate variability by applying nonlinear methods with different approaches for graphical representation of results. *Analysis, 6*(8), 38–45.

11. Xiang-yang, W., Wei-yi, L., Hong-ying, Y., Pan-pan, N., & Yong-wei, L. (2015). Invariant quaternion radial harmonic Fourier moments for color image retrieval. *Optics & Laser Technology, 66*, 78–88.

12. Chen, B., Shu, H., Coatrieux, G., Chen, G., Sun, X., & Coatrieux, J. L. (2015). Color image analysis by quaternion-type moments. *Journal of Mathematical Imaging and Vision, 51*(1), 124–144.

13. Pal, G., Acharjee, S., Rudrapaul, D., Ashour, A. S., & Dey, N. (2015). Video segmentation using minimum ratio similarity measurement. *International Journal of Image Mining, 1*(1), 87–110.

14. Chaki, J., & Parekh, R. (2011). Plant leaf recognition using shape based features and neural network classifiers. *International Journal of Advanced Computer Science and Applications, 2*(10), 41–47.

15. Virmani, J., Dey, N., & Kumar, V. (2016). PCA-PNN and PCA-SVM based CAD systems for breast density classification. In *Applications of Intelligent Optimization in Biology and Medicine*(pp. 159–180). Springer, Cham, Switzerland.

7

Scale-Space Shape Features

Scale-space theory is a multi-scale signal representation paradigm formed by groups of machine learning, image processing and computer vision with complementary intentions from biological vision and physics [1]. It is a formal theory for managing image structures on various scales, portraying an image as a single parameter family of smooth images, the representation of scale and space, parameterized by the size of the smoothing kernel utilized to suppress fine structures.

7.1 Curvature Scale Space (CSS)

This method is created on multi-scale depiction and curvature to signify planar curves. The creation of a CSS image involves stages as follows [2]:

- The edge of the image is attained by utilizing the Canny edge detector.
- Then the Gaussian kernel is convolved with the edge of the image. This procedure is called smoothing of the image curve.
  - The convolution is comprised of two constraints: l is termed the arc length constraint and σ is called the scale constraint.
  - The scale constraint σ is slowly amplified or increases through convolution. The image become smoothed as σ rises.
  - When $\sigma = 0$, the smoothed image is identical to the original image with every detail and as the σ rises, smoothed image becomes blurred and the resultant image is produced with less details.
- The concluding CSS image is comprised of several arch shaped edges that hinge on the concavity (inflections or shape) of the object.
- The smoothing of the contour stops when the number of curvature zero crossings turn out to be zero.

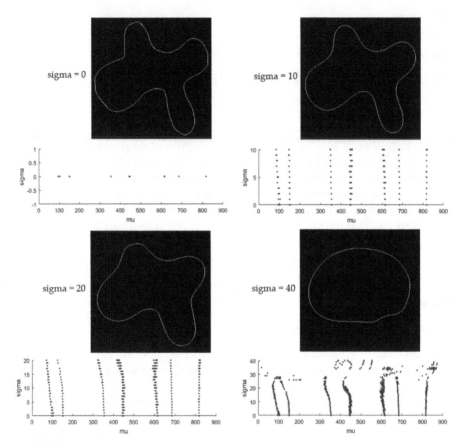

FIGURE 7.1
Steps of smoothing of a contour and CSS plots.

Figure 7.1 displays the stages in the smoothing of the specified image edge.

The CSS image is a binary image that signifies the edges of the image at several scale values [3]. For a continuous curve Γ, a parametric vector calculation can be represented by equation (7.1).

$$\Gamma(p) = (x(p), y(p)) \tag{7.1}$$

where p is the arc length of the contour. The expression for calculating the curvature function can be represented by equation (7.2).

$$K(p) = \frac{\dot{x}(p)\ddot{y}(p) - \ddot{x}(p)\dot{y}(p)}{\left(\dot{x}(p)^2 + \dot{y}(p)^2\right)^{3/2}} \tag{7.2}$$

The components ($X(\sigma,\mu)$ and $Y(\sigma,\mu)$) of the evolved curve of one-dimensional Gaussian kernel ($G(\sigma,\mu)$) with width p, is expressed by using equation (7.3).

$$X(\mu,\sigma) = x(p) * G(\mu,\sigma)$$
$$Y(\mu,\sigma) = y(p) * G(\mu,\sigma)$$

(7.3)

where * designates convolution. Conferring to the features of convolution, the derivatives of each component can be computed simply by using equation (7.4).

$$X_l(\mu,\sigma) = x(p) * G_l(\mu,\sigma)$$
$$X_{ll}(\mu,\sigma) = x(p) * G_{ll}(\mu,\sigma)$$
$$Y_l(\mu,\sigma) = y(p) * G_l(\mu,\sigma)$$
$$Y_{ll}(\mu,\sigma) = y(p) * G_{ll}(\mu,\sigma)$$

(7.4)

Now equation (7.2) can be represented as shown in equation (7.5).

$$K(\mu,\sigma) = \frac{X_l(\mu,\sigma)Y_{ll}(\mu,\sigma) - X_{ll}(\mu,\sigma)Y_l(\mu,\sigma)}{\left(X_l(\mu,\sigma)^2 + Y_l(\mu,\sigma)^2\right)^{3/2}}$$

(7.5)

Though for the original curve Γ, p is the standardized arc length constraint, the constraint l is not the standardized arc length for the changed curve Γ_σ. If the curvature zero crossings of Γ_σ are computed throughout evolution, they can be shown as points in the (μ,σ) plane. Here l approximates the standardized arc length and p of the Gaussian kernel. For each p there is a definite curve Γ_σ that comprises some curvature zero crossing points [4]. As p rises, Γ_σ turns out to be the number of zero crossings reduces and the curve becomes smoother. When the value of p turns adequately large, Γ_σ will become a convex curve with no curvature zero crossing, and the procedure of evolution or smoothing can be ended. The outcome of this procedure is the binary CSS image of the specified curve. The minor waves on the edge of the object are represented by the small contours of the CSS image and can be overlooked. Each contour is then characterized by the positions of the major maxima of its CSS image contours [5].

Though CSS has lots of advantages, it doesn't produce outcomes in conformance with a human apparition system at all times. The foremost weakness of this procedure is the issue of deep and shallow convexities/concavities of a shape. With the intention of overcoming these shortcomings, several variants of CSS are introduced.

7.1.1 Extreme Curvature Scale Space (ECSS)

This shape descriptor is used with the purpose of solving the issue of shallow and deep concavities of typical CSS. Whereas CSS utilizes curvature zero crossing values, the ECSS map is formed by chasing the location of maximum curvature points. ECSS is created on the maxima of the attained ECSS map [6]. It is vigorous with regard to scale, noise and orientation variations of the shape. The ECSS has proven to be an effective shape descriptor when equated to the typical CSS, specifically regarding deep or shallow concavities.

ECSS is used to prevail over the shortcomings of typical CSS. A group of multi-scale curvatures K (μ, σ) which resemble the group of curves of the shape can be defined as shown in equation (7.6).

$$K(\mu,\sigma) = \frac{x_u(\mu,\sigma)\,y_{uu}(\mu,\sigma) - x_{uu}(\mu,\sigma)\,y_u(\mu,\sigma)}{\left(x_u(\mu,\sigma)^2 + y_u(\mu,\sigma)^2\right)^{3/2}} \tag{7.6}$$

where x_u, y_u, x_{uu}, y_{uu} are correspondingly the first and second derivatives of x and y with regard to "u." This denotes extrema curvature zero crossing points. As σ rises, the inflection points reduce and consequently the inflection point vanishes at maximum scale value. The algorithm terminates when all the curvature zero-crossing points vanish [7]. With all maximum curvature points from the ECSS image the concluding ECSS descriptor contour is formed. Then the peaks (i.e., the maxima) are extracted and sorted. The ECSS descriptors are composed of all extreme points in the ECSS image.

Therefore, this technique offers two modifications to the standard CSS: for curvature zero crossings it replaces the utilization of contour extrema and enhances the curvature value in every extreme as an extra feature on each of the corresponding points. One of the big advantages is to assimilate the curvature feature for the descriptor, to differentiate among shape with deep and shallow concavities.

7.1.2 Direct Curvature Scale Space (DCSS)

DCSS is one of the effective variations of standard CSS which is utilized to detect corners [7]. DCSS is demarcated as the CSS that results from convolving a Gaussian kernel with the curvature of a planar curve directly.

Let $\varphi(p)$ be the function of the planar curve and $G(p,\sigma)$ be the Gaussian kernel. The curvature of the curve is denoted as $K(p) = \varphi(p)$. The DCSS of the image is created by convolving $K(p)$ directly with $G(p,\sigma)$. The resultant curvature function (convolved) $K(p,\sigma)$ is represented by $K(p,\sigma) = K(p) * G(p,\sigma)$ and can be represented by equation (7.7).

$$K(p,\sigma) = \int\limits_{-\infty}^{+\infty} K(l)G(p-l,\sigma)dl = \frac{1}{\sigma\sqrt{2\pi}} \int\limits_{-\infty}^{+\infty} K(l)e^{\frac{(p-l)^2}{2\sigma^2}} dl \qquad (7.7)$$

To compute corners at a specified scale σ, all the positions which have maxima absolute curvature, $|K(p,\sigma)|$, as well as the positive maxima and negative minima have to be resolved. A DCSS image is then created using $\max_{p,\sigma}|K(p,\sigma)|$, where p signifies the contour arc length. The DCSS image of a specified curve delivers the corner position information at variable scales. There are two representations that can be utilized to examine the features of the DCSS image [8]. The single corner termed the Γ model is utilized to demonstrate the reliable way of the line form behavior in the DCSS image. The double corner, for instance STAIR and END model, demonstrates the communication among a couple of adjacent corners with regard to scale. However, DCSS is delicate to noise; the hybrid method of DCSS/CSS can be efficiently utilized to overcome it.

7.1.3 Affine Resilient Curvature Scale Space (ARCSS)

Corner detectors based on CSS utilize arc-length as parameterized curvature. Thus, they aren't vigorous to affine alterations as the curve arc length doesn't persist invariant under affine alterations. The curve affine-length is comparatively invariant to affine alterations. Therefore, affine resilient corner detectors are used [9]. In this method, to bring out the edges from the grey scale images, canny edge detector is utilized. The local maxima of absolute curvature is recognized as a corner. Wrong corners are removed by equating every curvature maximum with its two adjacent minima built on the hypothesis that the corner point curvature should be at least twice the curvature of an adjacent minimum. The process which is utilized in ARCSS is described as follows [10]:

1. Discover the edge image utilizing the canny edge detector.
2. Extract edges from the image.
   a. Fill openings (if any) that are in a range and choose elongated edges.
   b. T-corners can be identified from T-junctions.
3. Parameterize every edge with its affine-length.
4. Absolute curvature is calculated for every parameterized edge at a suitable scale, and corners are identified by equating the curvature maxima to the equivalent edge curvature threshold in the adjacent minima.
5. Track the corners down to the lowermost scale seeing a small neighborhood to enhance localization.
6. Eliminate multiple existences of identical corners.

ARCSS corner detection is superior to CSS and ECSS corner detection. While standard CSS loses some accurate corners and ECSS is acquainted with some weak corners, ARCSS detectors deletes few true corners and acquaint with few weak corners. This is because the ARCSS detector utilizes the affine-length parameterization, but the standard CSS and ECSS detectors utilize the arc-length parameterization. Correspondingly, ARCSS is more resistant to noise as compared to its counterparts.

There are some advantages and limitations of CSS that are designated as follows:

Advantages: CSS is robust with regard to noise, scale and variation in orientation. Furthermore, both feature extraction and shape similarity matching are done very quickly. It computes the key features of a shape, permitting similarity-based retrieval. It is reliable, fast and compact. The local information of a shape is also preserved.

Limitations: The main limitation of CSS is the existence of uncertainties in CSS matching because of the issue of shallow concavities of the shape. It can be revealed that the deep and shallow concavities may generate the same big contours on the CSS image. Thus, a deep concavity may be matched with a shallow one during the CSS matching. Also, CSS is not suitable to open curves.

7.2 Morphological Scale Space (MSS)

MSS is a multi-scale boundary depiction constructed on morphological operations [11]. An image boundary is gradually smoothed by a number of opening and closing operations utilizing a group of structuring elements of growing size, producing a multiple scale depiction of the object. Then, equivalent features of the smoothed boundary across a range of scales are extracted and connected together establishing a map called the morphological scale space (MSS).

The basic morphological operators utilized for the explanation of MSS are as follows:

1. The erosion of the image I by the structuring element S is demarcated by equation (7.8).

$$I \ominus S = \bigcap_{s \in S} I_{-s}$$

(7.8)

where I_{-s} is the translation of the image I by $-s$. Figure 7.2 shows the erosion output of an image.

Original Image

Binary Image

Erosion Output

FIGURE 7.2
Erosion output of an image.

Original Image

Binary Image

Dilation Output

FIGURE 7.3
Dilation output of an image.

2. The dilation of the image I by the structuring element S is demarcated by equation (7.9). Figure 7.3 shows the dilation output of an image.

$$I \oplus S = \bigcup_{s \in S} I_{+s} \tag{7.9}$$

3. The opening of the image I by the structuring element S is demarcated by equation (7.10). Figure 7.4 shows the opening output of an image.

$$I \circ S = (I \ominus S) \oplus S \tag{7.10}$$

4. The closing of the image I by the structuring element S is demarcated by equation (7.11). Figure 7.5 shows the closing output of an image.

$$I \bullet S = (I \oplus S) \ominus S \tag{7.11}$$

It can be seen from Figures 7.2 through 7.5 that erosion shrinks the pattern, while dilation increases the pattern. Opening suppresses sharp lumps and removes narrow channels, while closing fills in gaps and tiny holes.

Original Image

Binary Image

Opening Output

FIGURE 7.4
Opening output of an image.

FIGURE 7.5
Closing output of an image.

MSS is represented using smooth boundary segments [12]. Boundary smoothness can be represented by equation (7.12).

$$\left(I \circ S(r)\right) \cap \left(I \bullet S(r)\right) \tag{7.12}$$

The first term (opening) eliminates corners whereas the second term (closing) fills in holes and cracks. The subsequent boundary points that are common to both operations belong to smooth segments of the boundary, therefore their intersection must contain smooth points [13]. This method when applied to a shape at several scales (r) would deliver a smoothness scale function, i.e., the MSS map. By filtering an image with scaled structuring elements, scale-spaces can be formed.

Figure 7.6 shows the smoothing of contour with MSS.

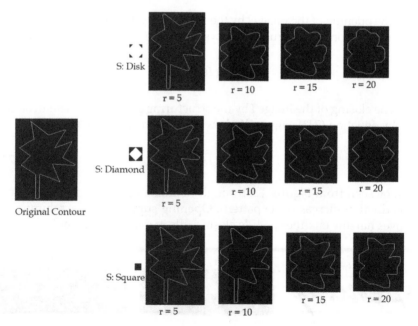

FIGURE 7.6
Smoothing of 2D contour by different structuring element.

There are some advantages and drawbacks of the MSS approach.

Advantages: MSS has a contour preservation property. The contours (i.e., the transitions of luminance) must be preserved: if a region is not removed, then it has to be not degraded.

Limitations: Edges of objects become very blurred at high scales.

7.3 Intersection Points Map (IPM)

IPM utilizes the intersection points among the original and the smoothed curve [14]. As the Gaussian kernel's standard deviation (σ) of the curve rises, the total number of the intersection points declines. You can determine the pattern functions by analyzing these residual points. Since this technique only works with the smoothing of the curve, only the convolution process is necessary to obtain the smooth curve. Thus, this technique is faster as compared to CSS for similar performances.

Figure 7.7 demonstrates an instance of IPM. The first row denotes the original images and the second row shows the intersected points before (left) and after (right) smoothing.

Intersected Points: 16   Intersected Points: 8

FIGURE 7.7
IPM illustration.

There are some advantages and limitations of IPM that are listed as follows:

Advantages: The IPM pattern can be recognized irrespective of its translation, rotation and scale variation. It is also robust to noise for a noise range.

Limitations: The chief drawback of this method is that it can't tackle the contours having suffered a non-rigid distortion and those which are occulted.

7.4 Summary

The scale-space concept can acquire plentiful information about an edge with various scales. In scale-space, from larger scales the global information of the pattern can be understood, whereas from lesser scales the detailed information of the pattern can be understood. Scale-space procedure is advantageous as the shape is made less delicate to errors in the arrangement or edge extraction procedures. The robustness to noise and the beautiful consistency with human observation are the benefits.

References

1. Mokhtarian, F. (1995). Silhouette-based isolated object recognition through curvature scale space. *IEEE Transactions on Pattern Analysis and Machine Intelligence,* *17*(5), 539–544.
2. Konur, U., Gürgen, F. S., Varol, F., & Akarun, L. (2015). Computer aided detection of spina bifida using nearest neighbor classification with curvature scale space features of fetal skulls extracted from ultrasound images. *Knowledge-Based Systems, 85,* 80–95.
3. Yang, J., Wang, H., Yuan, J., Li, Y., & Liu, J. (2016). Invariant multi-scale descriptor for shape representation, matching and retrieval. *Computer Vision and Image Understanding, 145,* 43–58.
4. Wang, P., Hu, X., Li, Y., Liu, Q., & Zhu, X. (2016). Automatic cell nuclei segmentation and classification of breast cancer histopathology images. *Signal Processing, 122,* 1–13.
5. Larsen, A. B. L., Vestergaard, J. S., & Larsen, R. (2014). HEp-2 cell classification using shape index histograms with donut-shaped spatial pooling. *IEEE Transactions on Medical Imaging, 33*(7), 1573–1580.

6. Herdiyeni, Y., Lubis, D. I., & Douady, S. (2015, November). Leaf shape identification of medicinal leaves using curvilinear shape descriptor. In *Soft Computing and Pattern Recognition (SoCPaR), 2015 7th International Conference of* (pp. 218–223). IEEE.

7. Zhong, B., & Liao, W. (2007). Direct curvature scale space: Theory and corner detection. *IEEE Transactions on Pattern Analysis and Machine Intelligence, 29*(3), 508–512.

8. Kopf, S., Haenselmann, T., & Effelsberg, W. (2005, July). Enhancing curvature scale space features for robust shape classification. In *Multimedia and Expo, 2005. ICME 2005. IEEE International Conference on* (p. 4). IEEE.

9. Awrangjeb, M., Lu, G., & Murshed, M. (2007, April). An affine resilient curvature scale-space corner detector. In *Acoustics, Speech and Signal Processing, 2007. ICASSP 2007. IEEE International Conference on* (Vol. 1, pp. I–1233). IEEE.

10. Awrangjeb, M., & Lu, G. (2008). An improved curvature scale-space corner detector and a robust corner matching approach for transformed image identification. *IEEE Transactions on Image Processing, 17*(12), 2425–2441.

11. Chaki, J., & Dey, N. (2018). *A Beginner's Guide to Image Preprocessing Techniques.* CRC Press, Boca Raton, FL.

12. Tian, Z., Dey, N., Ashour, A. S., McCauley, P., & Shi, F. (2017). Morphological segmenting and neighborhood pixel-based locality preserving projection on brain fMRI dataset for semantic feature extraction: an affective computing study. *Neural Computing and Applications, 30*, 1–16.

13. Wang, Y., Shi, F., Cao, L., Dey, N., Wu, Q., Ashour, A. S., Sherratt, S., Rajinikanth, V., & Wu, L. (2018). Morphological segmentation analysis and texture-based support vector machines classification on mice liver fibrosis microscopic images. *Current Bioinformatics, 14*(4), 282–294.

14. Zhang, Y. (2017). *Image Analysis.* Walter de Gruyter GmbH & Co KG, Berlin, Germany.

8

Shape Transform Domain Shape Feature

A given image signal can be converted between the time and frequency domains with a pair of mathematical operators called transforms [1]. One of the main purposes of transformation is to simplify mathematical analyses. For mathematical systems governed by linear differential equations, a very important class of systems with many real-world applications, converting the system description from the time domain to a frequency domain makes the differential equations much easier to solve. The shape descriptor is signified by all or part of transformed coefficients.

8.1 Fourier Descriptors

This is an effective shape explanation tool [2]. The shape explanation and classification utilizing a Fourier descriptor either in regions or boundary are easy to calculate, compact and vigorous to noise. It has several applications in various fields.

8.1.1 One-Dimensional Fourier Descriptors

This is attained by executing Fourier transform (FT) onto the signature of the shape, which is a 1D function that results from shape contour coordinates [3]. The standardized FT coefficients are called the shape's Fourier descriptor. A Fourier descriptor obtained from various signatures has substantial diverse performance on shape retrieval. A Fourier descriptor resulting from centroid distance function $D(t)$ outperforms a Fourier descriptor obtained from other various signatures of the shape in complete presentation. The discrete FT of $D(t)$ is then specified by equation (8.1).

$$F_m = \frac{1}{M} \sum_{t=0}^{M-1} D(t) \exp\left(\frac{-j2\pi mt}{M}\right), m = 0, 1, \ldots, M-1, j = \sqrt{-1} \qquad (8.1)$$

As the $D(t)$ is not variant to only translation and orientation, the attained Fourier coefficients (FC) need further regularized to become scaling and initial point independent [4]. The universal representation of FC of $D(t)$ altered through scaling and alteration of initial point from the original function $D(t)^{(0)}$ is represented by equation (8.2).

$$F_m = \exp(jm\alpha) \cdot p \cdot F_m^{(0)} \tag{8.2}$$

where $F_m^{(0)}$ and F_m are the FC of the original and transformed shape, respectively; α is the angle acquired by the alteration of initial point; and p is the scale factor. Normalized FC is represented by equation (8.3).

$$NF_m = \frac{F_m}{F_1} = \frac{\exp(jm\alpha) \cdot p \cdot F_m^{(0)}}{\exp(j\alpha) \cdot p \cdot F_1^{(0)}} = \frac{F_m^{(0)}}{F_1^{(0)}} \exp\big[j(m-1)\alpha\big] = NF_m^{(0)} \exp\big[j(m-1)\alpha\big] \tag{8.3}$$

where $NF_m^{(0)}$ and NF_m are the normalized FC of the original and transformed shape, respectively. If only magnitude of the coefficients is used and the phase information is ignored, then $|NF_m|$ and $NF_m^{(0)}$ are identical. Hence, $|NF_m|$ is invariant to orientation, scaling, translation and alteration of initial point.

The group of magnitudes of the shape's normalized FC $\{\,|NF_m|\,, 0 < m < M\}$ are utilized as shape descriptors. The one-dimensional Fourier descriptor has various good features such as easy derivation, easy normalization and ease of matching.

8.1.2 Region-Based Fourier Descriptor

This descriptor is denoted as generic Fourier descriptor (GFD) and can be utilized for wide-ranging applications [5]. Fundamentally, GFD is obtained by applying an improved polar FT on a shape image. To apply modified polar FT, the polar shape image is considered as a standard rectangular image. The steps are as described as follows:

- *Step 1:* The approximated normalized image is oriented in anti-clockwise direction by a small angular step.
- *Step 2:* Starting from the image centroid, the pixel values along positive x-direction are pasted into another new row matrix.
- Steps 1 and 2 are repeated until the image is oriented by 360°.

The outcome of the above-mentioned steps is a polar space image which is plotted into Cartesian space. The FT is attained by executing a discrete two-dimensional FT on the image shape as given by equation (8.4).

$$PF(r,\theta) = \sum_{\rho}\sum_{i} F(\rho,\phi_i) \exp\left[j2\pi\left(\frac{\rho}{R}r\right) + \frac{2\pi i}{T}\theta\right], 0 \le r < R, 0 \le \theta < T \quad (8.4)$$

where $0 \le \rho = \sqrt{(x - x_c)^2 + (y - y_c)^2} < R, \phi_i = i\left(\frac{2\pi}{T}\right)$

where (x_c, y_c) is the centroid of the shape, and T and R are the angular and radial resolutions. The attained Fourier coefficients are translation invariant. Scaling and rotation invariant generic Fourier coefficients (GF) are attained by equation (8.5).

$$GF = \left\{ \frac{|PF(0,0)|}{A}, \frac{|PF(0,1)|}{|PF(0,0)|}, \cdots, \frac{|PF(0,m)|}{|PF(0,0)|}, \cdots, \frac{|PF(0,n)|}{|PF(0,0)|}, \cdots, \frac{|PF(n,m)|}{|PF(0,0)|} \right\} \quad (8.5)$$

where A is the bounding circle's area inside which the polar image exists, n is the utmost number of the chosen radial frequencies, and m is the utmost number of chosen angular frequencies. The values n and m can be attuned to attain a hierarchical coarse to fine depiction requirement.

Generic Fourier coefficients are rotation, scale and translation invariant as shown in Figure 8.1.

There are some advantages and limitations of Fourier transform which are listed as follows:

Advantages: Fourier transform possesses an exclusive place in the analysis of several linear operators, basically because the complex exponentials are the eigenfunctions/eigenvectors of linear, shift-invariant operators. Fourier transform leads to an enormously influential theory of smoothness, due to the correspondence among decay and differentiability of the Fourier coefficients.

Limitations: Fourier transform delivers no information on the temporal/ spatial localization of features. A Fourier transform can express that there *is* a discontinuity, but it can't say *where* it is.

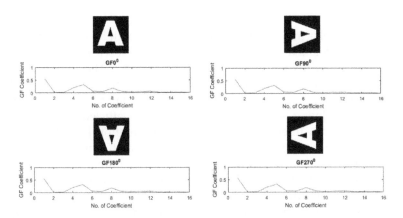

FIGURE 8.1
Generic Fourier coefficients of images.

8.2 Wavelet Transform

Wavelet transform analyzes the image signal by multiplying it by a window function and performing an orthogonal expansion [6]. This descriptor has several required features such as multi-resolution presentation, individuality, constancy, spatial localization and invariance.

In wavelet analysis, a scalable modulated window is shifted along the signal and for each location the spectrum is computed. A scalable modulated window, also recognized as mother of the wavelet, is the key point in which wavelet can be utilized to characterize breaks or sharp spikes of a signal. The transform is demarcated by a series of wavelet coefficients that are produced utilizing the wavelet mother function as shown in equation (8.6).

$$\psi_{i,j}(n) = 2^{-i/2}\psi\left(2^{-i}n - j\right) \tag{8.6}$$

where i is scaling constraint, j is a shifting constraint, and $i^{-1/2}$ is for energy normalization across various scales. The discrete wavelet transforms of function $f(n)$ are then calculated by utilizing equation (8.7).

$$W(i,j) = \sum_{n \in Z} f(n)\psi_{i,j}(n) \tag{8.7}$$

Inverse discrete wavelet transforms are represented by equation (8.8).

$$f(n) = \sum_{i}\sum_{j} W(i,j)\psi_{i,j}(n) \tag{8.8}$$

Practically, wavelet transform should be applied with a fast algorithm such as multi-resolution analysis. This transform decomposes a signal into a hierarchical tree of approximation and details coefficients. At every level i, a discrete wavelet transform generates approximation coefficients (wA_i) and detail coefficients (wD_i). Approximation coefficients are attained by convolving the signal with the low-pass filter, while the detail coefficients are convolved with the high-pass filter, trailed by dyadic decimation [7]. The high-pass and low-pass filter which are used to generate detail (D) and approximation (A) wavelet coefficients are shown in Figure 8.2, and the multi-level discrete wavelet decomposition tree and reassembling of the coefficients to form the original signal are shown in Figure 8.3.

Figure 8.4 shows the commonly used mother wavelets for wavelet decomposition.

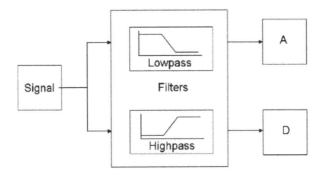

FIGURE 8.2
High-pass and low-pass filter that is used to generate detail (D) and approximation (A) wavelet coefficients.

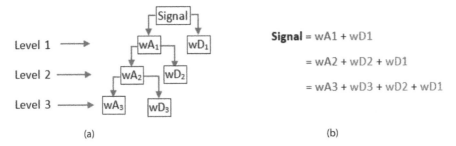

(a)   (b)

FIGURE 8.3
(a) Multi-level wavelet decomposition tree; (b) Reassembling original signal.

Similarly, discrete wavelet decomposition and reconstruction were created for two-dimensional signals or images [8]. The two-dimensional discrete wavelet transform decomposes an image at level i into four components (one approximation coefficient and three detail coefficient) at level $i + 1$:

1. Approximation coefficient wA_{i+1},
2. Horizontal detail coefficient wD_{i+1}^h,
3. Vertical detail coefficient wD_{i+1}^v and
4. Diagonal detail coefficient wD_{i+1}^d.

The two-dimensional discrete wavelet decomposition step is shown in Figure 8.5.

The symbol (Column↓2) denotes down-sampling columns by preserving only even indexed columns. Correspondingly, (Row↓2) represents down-sampling rows by preserving only evenly indexed rows. The pictorial representation of wavelet decomposition of an image of dimension $P \times Q$ is shown in Figure 8.6. A_i, HD_i, VD_i and DD_i denote the approximated

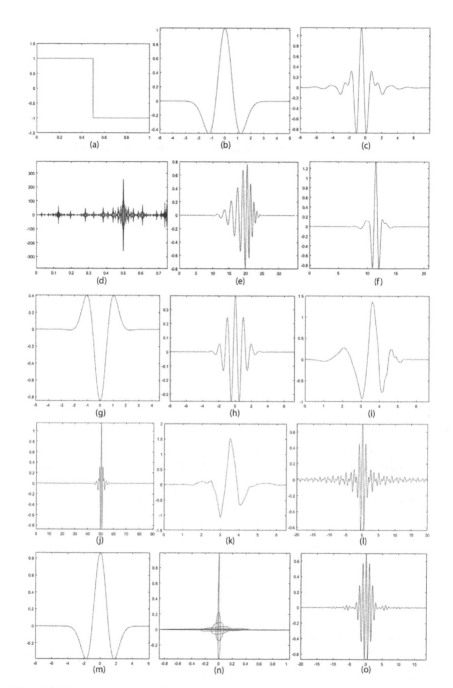

FIGURE 8.4
(a) Haar; (b) Gaussian; (c) Meyer; (d) Biorthogonal; (e) Daubechies20; (f) Coiflets; (g) Complex Gaussian; (h) Complex Morlet; (i) Daubechies4; (j) Discrete approximation of Meyer; (k) Symlets; (l) Shannon; (m) Mexican hat; (n) Fejer-Korovkin; (o) Frequency B-spline.

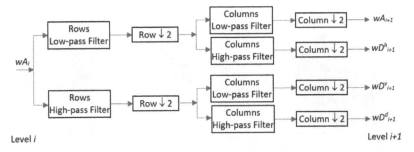

FIGURE 8.5
The two-dimensional discrete wavelet decomposition step.

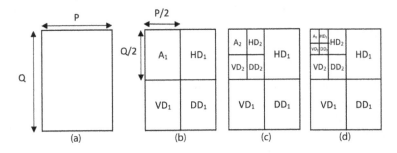

FIGURE 8.6
Wavelet decomposition of an image. (a) Original image; (b) First-level decomposition; (c) Second-level decomposition; (d) Third-level decomposition.

coefficient, horizontal detail coefficient, vertical detail coefficient and diagonal detail coefficient, respectively. i represents the level of decomposition.

Figure 8.7 shows the two-dimensional wavelet decomposition at level 1.

This feature is not variant to rotation, scaling and translation. The equivalence or matching procedure using this method can be completed cheaply.

There are some advantages and limitations of wavelet decomposition which are specified as follows:

Advantages: The foremost advantages of wavelet methods over traditional Fourier approaches are the utilization of localized basis functions and the quicker computation speed. Localized basis functions are perfect for analyzing real physical circumstances in which a signal comprises discontinuities and sharp points [9]. Wavelet transforms, regardless of having irregular shape, are able to effortlessly reconstruct functions with linear and higher order polynomial shapes, for instance, triangle, rectangle, second order polynomials, etc. Wavelets are able to denoise the specific signals far better than conventional filters that are based on Fourier transform design and that don't trail the algebraic rules followed by the wavelets.

Original Image (512 x 512)

A1 (256 x 256)　HD1 (256 x 256)

VD1 (256 x 256)　DD1 (256 x 256)

FIGURE 8.7
Level 1 wavelet decomposition.

> *Limitations:* Shift Sensitivity: The wavelet transform coefficients flop
> to differentiate among signal shifts. The down samplers in wavelet
> transform implementation are responsible for shift sensitivity. Poor
> Directionality: A two-dimensional transform undergoes poor direc-
> tionality if the transform coefficients are made known for only a few
> orientations like horizontal, vertical and diagonal details. Any image
> comprising smooth regions and edges with random orientation
> (natural images) can't be analyzed using standard two-dimensional
> wavelet transform as it exhibits poor directionality. Absence of Phase
> Information: The human visual system is sensitive to phase distortion
> formed by filtering the digital image with two-dimensional wavelet
> transform. So linear phase filtering can utilize symmetric extension
> approaches to sidestep the issue of bigger data size in image process-
> ing. As wavelet transform realization can't produce local phase infor-
> mation; complex-valued filtering is vital to avail local phase.

8.3 Angular Radial Transformation (ART)

ART is built in a polar coordinate system where the sinusoidal basis functions
are demarcated on a unit disk [10]. Let the image function in polar coordinate
be $I(r, \theta)$. Angular radial coefficient A_{mn} (m and n are the radial and angular
resolution) can be defined by equation (8.9).

$$A_{mn} = \int_0^{2\pi} \int_0^1 C_{mn}(r,\theta) I(r,\theta) r\,dr\,d\theta \tag{8.9}$$

$C_{mn}(r, \theta)$ is the angular radial transformation basis function and is divisible in the radial and angular direction so that the relation shown in equation (8.10) can be possible.

$$C_{mn}(r,\theta) = T_n(\theta)R_m(r) \qquad (8.10)$$

The angular basis function $T_n(\theta)$ is an exponential function utilized to eliminate rotation variance. This function is represented by equation (8.11).

$$T_n(\theta) = \frac{1}{2\pi}e^{jn\theta} \qquad (8.11)$$

The radial basis function $R_m(r)$ is denoted by equation (8.12).

$$R_m(r) = \begin{cases} 1 & if : m = 0 \\ 2\cos(\pi mr) & otherwise \end{cases} \qquad (8.12)$$

For scale normalization, the coefficients of ART are divided by the magnitude of angular radial transformation coefficient of order $m = 0$, $n = 0$.

Some advantages and limitations of angular radial transformation are discussed as follows:

Advantages: Angular radial transformation delivers a compact and effective technique to express pixel distribution in a two-dimensional object region. It can designate both disconnected and connected region shapes.

Limitations: Angular radial transformation is not robust to each rotation and to perspective projections. A planar object in a normal scene can be observed according to all orientations and agreed by an indefinite plan. This highly possible situation will interrupt the shape in the image and will foil the identification.

8.4 Shape Signature Harmonic Embedding

A harmonic function [11] is attained using a convolution among the Poisson kernel $P_K(r, \theta)$ and a given contour function $c(Ke^{j\varphi})$. Poisson kernel is denoted by equation (8.13).

$$P_K(r,\theta) = \frac{K^2 - r^2}{K^2 - 2Kr\cos(\theta) + r^2} \qquad (8.13)$$

The contour function can be any complex or real valued function. The harmonic function can be represented by equation (8.14).

$$c\left(re^{j\theta}\right) = \frac{1}{2\pi} \int_0^{2\pi} c\left(Ke^{j\varphi}\right) P_K\left(r,\varphi-\theta\right) d\varphi \tag{8.14}$$

The Poisson kernel $P_K(r,\theta)$ has a low-pass filter feature, where the radius r is in reverse association to the filter bandwidth. The radius r is measured as a scale constraint of a multi-scale depiction. One more vital property is $P_K(0, \theta) = 1$, representing $c(0)$ as the average value of contour function $c(Ke^{j\varphi})$.

The shape signature harmonic embedding depiction is invariant to translation, and it is invariant to rotation up to a horizontal shift in the rectilinear presentation. Also, it is invariant to unvarying to scaling. Shape signature harmonic embedding is vigorous to noise.

Some advantages and limitations of this method are listed as follows:

Advantages: It is computationally simple.

Limitations: Unfortunately, it is delicate to noise. Slight variations in contour cause huge errors in the matching process. It is not invariant to scale, translation and orientation. Usually further processing is needed to raise the robustness.

8.5 $\mathfrak{R}$-Transform

The $\mathfrak{R}$-transform to denote a shape is created on the Radon transform [12]. The method is described as follows. Assuming that the image shape function is $I(x, y)$, its Radon transform is demarcated by equation (8.15).

$$S_R\left(r,\theta\right) = \int_{-\infty}^{\infty} \int_{-\infty}^{\infty} I\left(x,y\right) \delta\left(x\cos\theta + y\sin\theta - r\right) dx dy \tag{8.15}$$

where δ is the Dirac delta function and represented by equation (8.16).

$$\delta\left(x\right) = \begin{cases} 1 & if : x = 0 \\ 0 & otherwise \end{cases} \tag{8.16}$$

$\theta \in [0, \pi]$ and $r \in (-\infty, \infty)$. Radon transform $S_R(r, \theta)$ is the integral of I over the line $L(r, \theta)$ demarcated by $r = x \cos \theta + y \sin \theta$.

Figure 8.8 is an illustration of a shape and its Radon transform.

$\mathfrak{R}$-transform is defined by equation (8.17).

$$\mathfrak{R}_I\left(\theta\right) = \int_{-\infty}^{\infty} S_R^2\left(r,\theta\right) dr \tag{8.17}$$

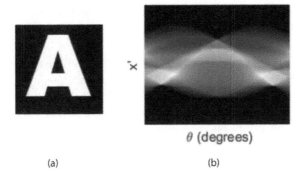

(a)   (b)

FIGURE 8.8
(a) Shape; (b) Radon transform of (a).

where $S_R(r, \theta)$ is the Radon transform of the domain shape function I. The properties of $\Re_I(\theta)$ are as follows:

- *Orientation*: The orientation of the image by an angle θ_0 indicates a translation of the $\Re$-transform of θ_0: $\Re_I(\theta + \theta_0)$.
- *Periodicity*: $\Re_I(\theta \pm \pi) = \Re_I(\theta)$.
- *Scaling*: A scale alteration of the shape I encourages a scaling in the amplitude of this transform.
- *Translation*: This method is translation invariant under a shape function I by a vector $m = (x_0, y_0)$.

Provided the shape's distance transform, the distance image is divided into M levels of similar distance to preserve the division isotropic. For every distance level, pixels consuming a distance value larger than that level are chosen and at every level of division, a $\Re$-transform is calculated. In this way, both the interior formation and the contours of the shape are taken.

Orientation of the shape infers an equivalent change of the $\Re$-transform [13]. Thus, a one-dimensional FT is executed on this function to acquire the invariant rotation. $\Re$-transform descriptor vector is demarcated by equation (8.18).

$$D = \left(\frac{F\Re^1\left(\dfrac{\pi}{N}\right)}{F\Re^1(0)}, \cdots, \frac{F\Re^1\left(\dfrac{i\pi}{N}\right)}{F\Re^1(0)}, \cdots, \frac{F\Re^1(\pi)}{F\Re^1(0)}, \cdots, \frac{F\Re^M\left(\dfrac{\pi}{N}\right)}{F\Re^M(0)}, \cdots, \frac{F\Re^M(\pi)}{F\Re^M(0)} \right)$$

$$(8.18)$$

where $i \in [1, N]$, F is discrete one-dimensional FT and N is the angular resolution. $F\Re^a$ is the magnitude of the FT to $\Re$-transform. $a \in [1, M]$ is the segmentation level of Chamfer distance transform.

There are some advantages and limitations of this method which are depicted as follows:

Advantages: The $\Re$-transform is invariant under scaling and translation. Its principle is simple.

Limitations: Specified a huge collection of shapes, a single $\Re$-transform for each shape is ineffective to discriminate from the others since the $\Re$-transform offers an extremely compact shape depiction. In this viewpoint, to enhance the description, every shape is mapped in the Radon space for various Chamfer distance transforms' segmentation levels. The $\Re$-transform compressed the form, so one $\Re$-transform per form doesn't describe the shape efficiently.

8.6 Shapelet Descriptor (SD)

SD is used to deliver an architecture to animate shapes and extract relevant slices of objects [14]. The architecture presumes that animate two-dimensional uncomplicated closed curve shapes are created by a linear placement of a quantity of shape bases. A basis function can be defined by $\psi(D; p, \sigma)$ where $p \in [0, 1]$ designates the position of the basis function comparative to the field of the detected curve, and σ is the scale of the function ψ. Figure 8.9 displays the function's shape using various σ values. It shows variation with various transforms and parameters.

The 2×2 matrix of basis coefficients as represented in equation (8.19) is utilized for the affine transformations of the basis function.

$$C_i = \begin{bmatrix} c_i & r_i \\ s_i & t_i \end{bmatrix} \tag{8.19}$$

The variables for denoting a base are represented by $A_i = (C_i, p_i, \sigma_i)$ and are called basis elements. The shapelet is denoted by equation (8.20).

$$\gamma(D; A_i) = C_i \psi(D; p_i, \sigma_i) \tag{8.20}$$

Figure 8.9b–d denote shapelets attained from the basis function ψ by the affine transformation of rotation (r_i), scaling (s_i) and shearing (t_i) respectively as shown in the basis coefficient C_i. By accumulating the entire shapelets at different p, σ, C and discretising them at various levels, a whole dictionary is attained as shown in equation (8.21).

$$\Delta = \{\gamma(D; A_i) : \forall A; b\gamma_0, b > 0\} \tag{8.21}$$

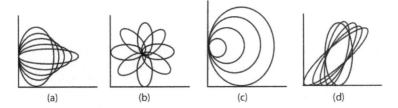

FIGURE 8.9
Lobe-shaped curve. (a) σ; (b) Rotation; (c) Scaling; (d) Shearing.

A special type of shapelet γ_0 is denoted as an ellipse [15]. Shapelets are construction blocks for shape boundaries and they generate closed curves by linear addition as represented by equation (8.22).

$$\Gamma(D) = \begin{bmatrix} x_0 \\ y_0 \end{bmatrix} + \sum_{i=1}^{M} \begin{bmatrix} c_i & r_i \\ s_i & t_i \end{bmatrix} \psi(D; p_i, \sigma_i) + R(D) \tag{8.22}$$

There are some advantages and limitations of shapelet descriptors that are listed as follows:

Advantages: Shapelets are comprehensible and can offer insight into the problem domain. The shapelet decomposition is mainly effective for images localized in space and offer a high level of compression.

Limitations: Shear estimations from circular shapelets are biased if the shape to be defined has too steep a profile or too large an ellipticity. Profile mismatch is the more vital source of bias. For elliptical shapelets, profile mismatch still poses a substantial problem, since the shapelet models can't fully detect biases of the ellipticity prior to when the profile becomes steeper than exponential.

8.7 Summary

As a type of global shape descriptor method, shape transform domain analysis receives the entire shape as the shape depiction. The description structure is modeled for this presentation. Unlike the study of spatial interrelations, shape projects transform a shape boundary or region into another field to achieve some of its basic characteristics. For a shape descriptor, there is always a balance between precision and efficacy. Shape should be defined as correct as possible, and a shape descriptor has to be as compressed as possible to facilitate indexing and recovery. It is easy to obtain a shape description with various levels of precision and efficacy by selecting the proper number of transform coefficients for a shape transformation analysis algorithm.

References

1. Borde, S., & Bhosle, U. (2012). Feature vectors based CBIR in spatial and transform domain. *International Journal of Computer Applications, 60*(19), 34–42.
2. Saba, L., Dey, N., Ashour, A. S., Samanta, S., Nath, S. S., Chakraborty, S., Sanches, J., Kumar, D., Marinho, R. T., & Suri, J. S. (2016). Automated stratification of liver disease in ultrasound: An online accurate feature classification paradigm. *Computer Methods and Programs in Biomedicine, 130*, 118–134.
3. Chaki, J., & Parekh, R. (2012). Plant leaf recognition using Gabor filter. *International Journal of Computer Applications, 56*(10), 26–29.
4. Roy, A. B., Dey, D., Mohanty, B., & Banerjee, D. (2012). Comparison of FFT, DCT, DWT, WHT compression techniques on electrocardiogram and photoplethysmography signals. In *IJCA Special Issue on International Conference on Computing, Communication and Sensor Network CCSN* (pp. 6–11).
5. Su, L., & Wu, H. T. (2017). Extract fetal ECG from single-lead abdominal ECG by de-shape short time Fourier transform and nonlocal median. *Frontiers in Applied Mathematics and Statistics, 3*, 2.
6. Dey, N., Roy, A. B., & Dey, S. (2012). A novel approach of color image hiding using RGB color planes and DWT. arXiv preprint arXiv:1208.0803.
7. Chaki, J., Parekh, R., & Bhattacharya, S. (2015). Plant leaf recognition using texture and shape features with neural classifiers. *Pattern Recognition Letters, 58*, 61–68.
8. Bhattacharya, T., Dey, N., & Chaudhuri, S. R. (2012). A session based multiple image hiding technique using DWT and DCT. arXiv preprint arXiv:1208.0950.
9. Dey, N., Nandi, P., Barman, N., Das, D., & Chakraborty, S. (2012). A comparative study between Moravec and Harris corner detection of noisy images using adaptive wavelet thresholding technique. arXiv preprint arXiv:1209.1558.
10. Amanatiadis, A., Kaburlasos, V. G., Gasteratos, A., & Papadakis, S. E. (2011). Evaluation of shape descriptors for shape-based image retrieval. *IET Image Processing, 5*(5), 493–499.
11. Maji, P., Chatterjee, S., Chakraborty, S., Kausar, N., Samanta, S., & Dey, N. (2015, March). Effect of Euler number as a feature in gender recognition system from offline handwritten signature using neural networks. In *Computing for Sustainable Global Development (INDIACom), 2015 2nd International Conference on* (pp. 1869–1873). IEEE.
12. Tabbone, S., Wendling, L., & Salmon, J. P. (2006). A new shape descriptor defined on the Radon transform. *Computer Vision and Image Understanding, 102*(1), 42–51.
13. Chaki, J., & Dey, N. (2018). *A Beginner's Guide to Image Preprocessing Techniques*. CRC Press, Boca Raton, FL.
14. Hore, S., Chakroborty, S., Ashour, A. S., Dey, N., Ashour, A. S., Sifaki-Pistolla, D., Bhattacharya, T., & Chaudhuri, S. R. (2015). Finding contours of hippocampus brain cell using microscopic image analysis. *Journal of Advanced Microscopy Research, 10*(2), 93–103.
15. Chaki, J., Parekh, R., & Bhattacharya, S. (2016). Plant leaf recognition using ridge filter and curvelet transform with neuro-fuzzy classifier. In *Proceedings of 3rd International Conference on Advanced Computing, Networking and Informatics* (pp. 37–44). Springer, New Delhi, India.

9

Applications of Shape Features

Shape is a vital visual and the emergent feature for image content explanation. The utilization of object shape is one of the most challenging problems in forming effective content-based image retrieval [1]. Shape content explanation can't be accomplished precisely since determining the similarity among shapes is problematic. Thus, two steps are important in shape-based image retrieval: extraction of the shape feature and similarity calculation among the extracted features. Some of the applications are mentioned in this chapter.

9.1 Digit Recognition

There are many ways of digit recognition using shape feature [2]. One of the algorithms for digit recognition using shape feature is as follows:

1. First draw digits from 0 to 9 with the help of paint software and save it in JPG format (Figure 9.1).
2. Divide images into training and testing set.
3. Convert all the images to binary [3], i.e., the contour pixels of the digit are 1s and the residual pixels are 0s (Figure 9.2).
4. From the binary image construct the chain code [4] using 8-connectivity. Figure 9.3 shows the chain code of Figure 9.2.
5. Use Euclidean distance measurement for comparison of chain codes between training and testing images [5].
6. The test digit is recognized by the minimum difference with the training images.

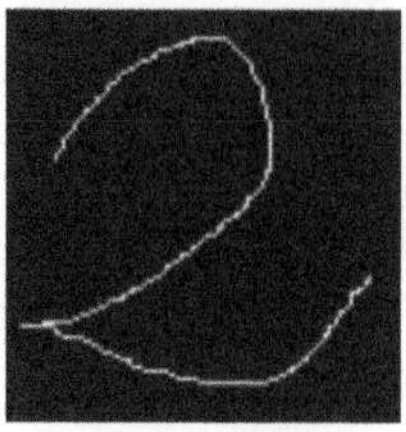

FIGURE 9.1
Digit two.

FIGURE 9.2
Binarized image.

```
0 0 0 0 0 7 0 0 0 0 0 0 0 7 0 7 0 0 7 0 7 0 7 7 0 0 7 0 7 7 0 0 7 7 0
7 0 7 7 6 0 7 0 7 6 0 0 7 6 0 7 7 7 0 0 7 6 7 7 0 7 7 0 7 6 7 6 0 7 7
7 6 7 6 6 7 6 6 6 6 6 6 6 6 6 6 6 6 5 6 6 6 5 6 6 6 6 5 5 6 5 6 6 5 5
5 5 5 5 5 4 4 4 4 4 4 4 4 3 4 4 3 4 4 3 4 4 4 3 2 4 3 4 4 3 3 4 3 4 3
2 3 4 3 2 3 3 3 3 3 3 3 3 3 3 2 3 3 3 2 2 3 3 2 6 7 7 6 6 7 7 7 6 7 7 7
6 0 7 7 6 0 7 7 6 7 0 7 6 7 0 7 0 7 7 0 0 7 7 7 0 0 0 7 0 7 0 0 7 0 0
0 0 0 0 0 0 0 1 1 1 1 1 1 1 2 2 1 2 1 1 2 2 2 2 1 2 2 2 1 2 2 2 2 2 2
2 2 2 2 2 2 2 3 2 3 2 3 3 3 3 2 3 3 3 4 3 4 3 2 3 3 4 4 3 3 3 3 3 4 3
3 4 3 3 3 3 4 3 4 3 3 4 3 4 3 3 4 4 4 3 3 4 3 4 3 4 4 3 4 3 4 4 4 4 3
1 1 1 0 0 0 0 0 1 0 1 0 1 1 0 0 1 0 1 0 0 2 1 0 0 1 0 0 0 0 0 0 1 0 1
0 0 0 0 0 1 0 0 0 0 0 0 0 0 0 0 0 0 0 0 0 0 0 0 0 7 0 0 7 0 7 7 0 7 7 7
7 7 7 7 7 6 7 6 0 7 6 7 6 0 7 7 6 6 6 0 7 0 7 6 7 2 3 3 3 4 3 2 2 3 3
3 3 2 3 3 2 3 3 3 3 3 3 2 4 3 3 3 4 3 4 3 4 4 3 4 4 4 4 4 4 4 4 4 4 4
4 4 4 4 4 4 4 5 4 4 4 4 4 5 4 5 4 4 4 4 4 5 4 4 5 5 4 5 4 5 4 4 5 5
4 5 4 5 4 4 4 4 4 5 4 6 5 4 4 4 4 4 4 4 4 0
```

FIGURE 9.3
The chain code of Figure 9.2.

9.2 Character Recognition

There are many methods of character recognition using shape feature [6]. One of the algorithms for character recognition using shape feature is as follows:

1. First draw characters from A to Z with the help of paint software and save it in JPG format (Figure 9.4).
2. Divide images into training and testing set.

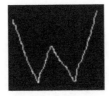

FIGURE 9.4
Character.

(a)   (b)   (c)   (d)

FIGURE 9.5
The wavelet coefficients of Figure 9.4. (a) Approximation coefficient; (b) Horizontal coefficient;
(c) Vertical coefficient; (d) Diagonal coefficient.

3. Generate the wavelet transform [7] from every image. Figure 9.5 shows the wavelet coefficients of Figure 9.4.

   From the coefficients it is clear that for the character "W," the vertical coefficient is more prominent among all the detail coefficients.

4. For comparison of wavelet coefficients between training and testing images, Euclidean distance measurement is used.

5. The test character is recognized by the minimum difference with the training images.

9.3 Fruit Recognition

There are many ways of fruit recognition using shape feature [8]. One of the algorithms for fruit recognition using shape feature is as follows:

1. First collect some images of different types of fruits that vary in shape. Figure 9.6 shows the image of an orange.

2. The images are divided into training and testing set.

3. Convert all the images to binary [9], i.e., the pixels of the fruit are 1s and the residual pixels are 0s (Figure 9.7).

FIGURE 9.6
Image of an orange.

FIGURE 9.7
Binarized image of Figure 9.6.

4. The fruit contour is then extracted using the canny edge detection algorithm [10] (Figure 9.8).

5. Calculate the centroid distance [11] for every image. Figure 9.9 shows the centroid distance plot of Figure 9.8.

6. For comparison of centroid distance between training and testing images, Euclidean distance measurement is used.

7. The test fruit image is recognized by the minimum difference with the training images.

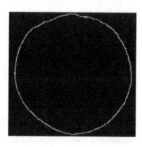

FIGURE 9.8
Contour of the fruit.

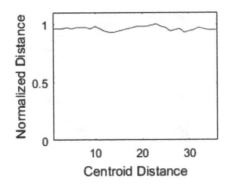

FIGURE 9.9
The centroid distance plot of Figure 9.8.

9.4 Leaf Recognition

There are many ways of leaf recognition using shape feature [12]. One of the algorithms for leaf recognition using shape feature is as follows:

1. First collect some images of different types of leaves that vary in shape. Figure 9.10 shows the image of a leaf.
2. The images are divided into training and testing set.
3. Convert all the images to binary, i.e., the pixels of the fruit are 1s and the residual pixels are 0s (Figure 9.11).
4. The leaf contour is then extracted using the canny edge detection algorithm (Figure 9.12).
5. Calculate the 7 Hu moment [13] for every image. Figure 9.13 shows the plot of 7 moment values of Figure 9.12.

FIGURE 9.10
Image of a leaf.

FIGURE 9.11
Binarized image of Figure 9.10.

FIGURE 9.12
Contour of the leaf.

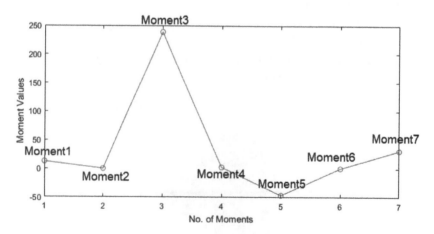

FIGURE 9.13
The Hu moment plot of Figure 9.12.

6. For comparison of moments between training and testing images, Euclidean distance measurement is used.

7. The test leaf image is recognized by the minimum difference with the training images.

9.5 Hand Gesture Recognition

There are many ways of hand gesture recognition using shape feature [14]. One of the algorithms for hand gesture recognition using shape feature is as follows:

1. First collect some images of hand gestures. Figure 9.14 shows the image of some collected hand gestures.

2. Convert the RGB image into grayscale and then into binary format (Figure 9.15).

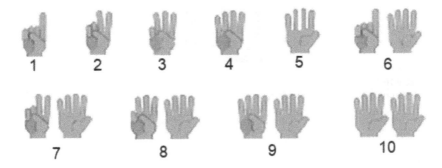

FIGURE 9.14
Hand gesture images.

FIGURE 9.15
Grayscale image (left) and binary image (right).

3. Image centroid is calculated using moment.

4. Detect peaks [15] of tip of finger.

5. Classify peaks as significant peaks by calculating the Euclidean distance between peak and centroid (Figure 9.16).

   If (distance > 70% of maximum peak distance)

   > Significant peak

   Else

   > Insignificant peak

6. Shape signature is generated with five binary sequence where 1 represents raised finger and 0 represents folded finger. The shape signature and binary code of Figure 9.16 is shown in Figure 9.17.

FIGURE 9.16
Significant peaks with centroid indicated with circles.

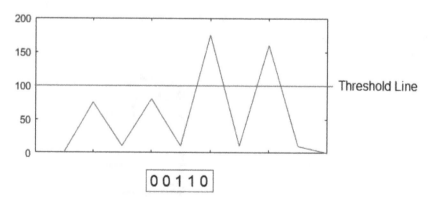

FIGURE 9.17
The shape signature of Figure 9.16.

9.6 Summary

The extraction of a form property in accordance with human observation is not a simple task. Because human perception and vision are an unusual, difficult system, we may naturally hope that machine vision performs brilliantly with minor complications. Furthermore, the selection of appropriate features for a shape recognition system must recognize which types of features are appropriate for the task. There is no overall shape function that works best for every type of image.

References

1. Zhang, D., & Lu, G. (2001, August). Content-based shape retrieval using different shape descriptors: A comparative study. In *Null* (p. 289). IEEE.
2. Belongie, S., Malik, J., & Puzicha, J. (2002). *Shape Matching and Object Recognition Using Shape Contexts*. California University of San Diego, Department of Computer Science and Engineering, San Diego, CA.
3. Chaki, J., & Dey, N. (2018). *A Beginner's Guide to Image Preprocessing Techniques*. CRC Press, Boca Raton, FL.
4. Shen, S., Bui, A. A., Cong, J., & Hsu, W. (2015). An automated lung segmentation approach using bidirectional chain codes to improve nodule detection accuracy. *Computers in Biology and Medicine*, *57*, 139–149.
5. Saha, M., Chaki, J., & Parekh, R. (2013). Fingerprint recognition using texture features. *International Journal of Science and Research (IJSR)*, *2*, 12.
6. Kamble, P. M., & Hegadi, R. S. (2015). Handwritten Marathi character recognition using R-HOG Feature. *Procedia Computer Science*, *45*, 266–274.
7. Dey, N., Mishra, G., Nandi, B., Pal, M., Das, A., & Chaudhuri, S. S. (2012, December). Wavelet based watermarked normal and abnormal heart sound identification using spectrogram analysis. In *Computational Intelligence & Computing Research (ICCIC), 2012 IEEE International Conference on* (pp. 1–7). IEEE.
8. Patel, H. N., Jain, R. K., & Joshi, M. V. (2012). Automatic segmentation and yield measurement of fruit using shape analysis. *International Journal of Computer Applications*, *45*(7), 19–24.
9. Chaki, J., Parekh, R., & Bhattacharya, S. (2015). Plant leaf recognition using texture and shape features with neural classifiers. *Pattern Recognition Letters*, *58*, 61–68.
10. Chaki, J. (2018, January). An efficient two-stage Palmprint recognition using Frangi-filter and 2-component partition method. In *2018 Fifth International Conference on Emerging Applications of Information Technology (EAIT)* (pp. 1–5). IEEE.

11. Chaki, J., & Parekh, R. (2011). Plant leaf recognition using shape based features and neural network classifiers. *International Journal of Advanced Computer Science and Applications*, 2(10), 41–47.
12. Chaki, J., Dey, N., Moraru, L., & Shi, F. (2019). Fragmented plant leaf recognition: Bag-of-features, fuzzy-color and edge-texture histogram descriptors with multi-layer perceptron. *Optik*, *181*, 639–650.
13. Chaki, J., & Parekh, R. (2017, December). Texture based coin recognition using multiple descriptors. In *2017 International Conference on Computer, Electrical & Communication Engineering (ICCECE)* (pp. 1–8). IEEE.
14. Stergiopoulou, E., & Papamarkos, N. (2009). Hand gesture recognition using a neural network shape fitting technique. *Engineering Applications of Artificial Intelligence*, *22*(8), 1141–1158.
15. Dey, N., Pal, M., & Das, A. (2012). A session based blind watermarking technique within the NROI of retinal fundus images for authentication using DWT, spread spectrum and harris corner detection. arXiv preprint arXiv:1209.0053.

Index